# Studies in Computational Intelligence

Volume 580

**Series editor**

Janusz Kacprzyk, Polish Academy of Sciences, Warsaw, Poland
e-mail: kacprzyk@ibspan.waw.pl

*About this Series*

The series "Studies in Computational Intelligence" (SCI) publishes new developments and advances in the various areas of computational intelligence—quickly and with a high quality. The intent is to cover the theory, applications, and design methods of computational intelligence, as embedded in the fields of engineering, computer science, physics and life sciences, as well as the methodologies behind them. The series contains monographs, lecture notes and edited volumes in computational intelligence spanning the areas of neural networks, connectionist systems, genetic algorithms, evolutionary computation, artificial intelligence, cellular automata, self-organizing systems, soft computing, fuzzy systems, and hybrid intelligent systems. Of particular value to both the contributors and the readership are the short publication timeframe and the world-wide distribution, which enable both wide and rapid dissemination of research output.

More information about this series at http://www.springer.com/series/7092

Stefka Fidanova

Editor

# Recent Advances in Computational Optimization

Results of the Workshop
on Computational Optimization
WCO 2013

 Springer

*Editor*
Stefka Fidanova
Department of Parallel Algorithms
Institute of Information and Communication
  Technology
Bulgarian Academy of Sciences
Sofia
Bulgaria

ISSN 1860-949X                    ISSN 1860-9503   (electronic)
Studies in Computational Intelligence
ISBN 978-3-319-12630-2          ISBN 978-3-319-12631-9   (eBook)
DOI 10.1007/978-3-319-12631-9

Library of Congress Control Number: 2014956569

Springer Cham Heidelberg New York Dordrecht London

Printed on acid-free paper

Springer International Publishing AG Switzerland is part of Springer Science+Business Media
(www.springer.com)

# Preface

Many real-world problems arising in engineering, economics, medicine, and other domains can be formulated as optimization tasks. Everyday we solve optimization problems. Optimization occurs in the minimizing time and cost or the maximization of the profit, quality, and efficiency. Such problems are frequently characterized by nonconvex, nondifferentiable, discontinuous, noisy or dynamic objective functions and constraints that ask for adequate computational methods.

This volume is a result of vivid and fruitful discussions held during the Workshop on Computational Optimization. The participants agree that the relevance of the conference topic and the quality of the contributions have clearly suggested that a more comprehensive collection of extended contributions devoted to the area would be very welcome and would certainly contribute to a wider exposure and proliferation of the field and ideas.

The volume includes important real problems like parameter settings for controlling processes in bioreactor, resource-constrained project scheduling, problems arising in transport services, error correcting codes, optimal system performance, energy consumption, and so on. Some of them can be solved applying traditional numerical methods, but others needs a huge amount of computational resources. Therefore, for them it is more appropriate to develop algorithms based on some metaheuristic method like evolutionary computation, ant colony optimization, constrain programming, etc.

April 2014                                                                 Stefka Fidanova

# Organization

Workshop on Computational Optimization (WCO 2013) is organized in the framework of Federated Conference on Computer Science and Information Systems FedCSIS—2013.

## Conference Co-chairs

| | |
|---|---|
| Stefka Fidanova | IICT (Bulgarian Academy of Sciences, Bulgaria) |
| Antonio Mucherino | IRISA (Rennes, France) |
| Daniela Zaharie | West University of Timisoara (Romania) |

## Program Committee

| | |
|---|---|
| Rumen Andonov | IRISA and University of Rennes 1, Rennes, France |
| David Bartl | University of Ostrava, Czech Republic |
| Janez Brest | University of Maribor, Slovenia |
| Douglas Goncalves | IRISA, University of Rennes 1, France |
| Hiroshi Hosobe | National Institute of Informatics, Japan |
| Hideaki Liduka | Kyushu Institute of Technology, Japan |
| Jouni Lampinen | University of Vaasa, Finland |
| Carlile Lavor | IMECC-UNICAMP, Campinas, Brazil |
| Pencho Marinov | Bulgarian Academy of Science, Bulgaria |
| Kaisa Miettinen | University of Jyvaskyla, Finland |
| Panos Pardalos | University of Florida, United States |
| Patrick Siarry | Universite Paris XII Val de Marne, France |
| Stefan Stefanov | Neofit Rilski University, Bulgaria |

Tomas Stuetzle            Universite Libre de Bruxelles, Belgium
Ponnuthurai Suganthan     Nanyang Technological University, Singapore
Josef Tvrdik              University of Ostrava, Czech Republic
Michael Vrahatis          University of Patras, Greece

# Contents

# A Three-Stage Heuristic for the Capacitated Vehicle Routing Problem with Time Windows

**Hakim Akeb, Adel Bouchakhchoukha and Mhand Hifi**

**Abstract** In this paper we propose to solve the Capacitated Vehicle Routing Problem with Time Windows (CVRPTW) by using a three-stage solution procedure. CVRPTW is defined by a set of customers (with associated demands and time windows) and a set of vehicles (with a given capacity) and the aim of the problem is to serve all the customers inside their respective time windows. There are two objectives to optimize, the first one consists in using the minimum number of vehicles and the second one is to minimize the total distance traveled by all the vehicles. The proposed heuristic combines (i) a clustering stage in which the set of customers is divided into disjoint clusters, (ii) a building stage that serves to provide a feasible tour in each cluster and, (iii) a local-search stage that is applied in order to try to improve the quality of the solutions obtained from the second stage. The computational investigation, conducted on a class of benchmark instances of the literature, shows that the results reached by the proposed heuristic remain competitive when compared to the best known solutions taken from the literature.

**Keywords** Beam search · Capacitated vehicle routing problem with time windows · Clustering · Local search

## 1 Introduction

The *Vehicle Routing Problem* (VRP) is one of the most studied problems of the routing problems family. The standard VRP consists to serve a set $N$ of $n$ customers by using a fleet composed of $m$ vehicles. The objective is to minimize the number

H. Akeb (✉)
ISC Paris Business School, 22 Boulevard du Fort de Vaux, 75017 Paris, France
e mail: hakeb@iscparis.com

A. Bouchakhchoukha
MSE Université Paris 1 Panthéon Sorbonne, 6–12 Bd de L'Hôpital, 75013 Paris, France

A. Bouchakhchoukha · M. Hifi
Université de Picardie Jules Verne, UR EPROAD, Équipe ROAD, 7 Rue du Moulin Neuf, 80039 Amiens, France

© Springer International Publishing Switzerland 2015
S. Fidanova (ed.), *Recent Advances in Computational Optimization*,
Studies in Computational Intelligence 580, DOI 10.1007/978-3-319-12631-9_1

of vehicles as well as the sum of distances traveled by these ones, each customer is visited exactly once. When a time window is associated with each customer, the problem becomes the *Vehicle Routing Problem with Time Windows* (VRPTW) and when a demand is considered for each customer the problem obtained is called the *Capacitated Vehicle Routing Problem with Time Windows* (CVRPTW).

In this paper, we propose a three-stage resolution approach-based heuristic for the CVRPTW. An instance of the CVRPTW is defined by the following parameters (these notations are those used in the rest of the paper):

- A set $N$ containing $n$ customers, where each customer (or *vertex*) $i \in N$ has coordinates $(x_i, y_i)$ in the Euclidean plan.
- Each customer $i \in N$ is characterized by its demand $d_i$, its time window $W_i = [e_i, l_i]$, and a service time $s_i$ denoting the time spent to serve the $i$th customer.
- A fleet $V$ containing $m$ identical vehicles, each one with a capacity $C_{\max}$.
- Each vehicle $v_j \in V$ starts from the *depot*, denoted by $D$ (with coordinates $(x_D, y_D)$), visits a subset of customers, and returns to the depot.
- Vehicle $v_j \in V$, $1 \leq j \leq m$, performs then a route $R_j$ (by visiting a given number of customers). The customers associated with route $R_j$ form cluster $C_j$.
- The sum of distances traveled by vehicle $v_j \in V$ is denoted by $\mathcal{D}_j$.
- The distance $dist_{ij} = \sqrt{(x_i - x_j)^2 + (y_i - y_j)^2}$ denotes the Euclidean distance between customers $i$ and $j$.

Then the CVRPTW can be formulated as follows:

$$\min\ m \tag{1}$$

$$\min \sum \mathcal{D}_j,\ 1 \leq j \leq m \tag{2}$$

*subject to*

$$\sum_{i \in C_j} d_i \leq C_{\max}, \quad \forall i \in C_j \tag{3}$$

$$t_i \in W_i = [e_i, l_i]\ \forall i \in N \cup \{D\} \tag{4}$$

Equation 1 denotes the first objective function that indicates the number of vehicles $m$ to minimize. Equation 2 is the second objective to minimize and corresponds to the total distance traveled by all vehicles. The first constraint (3) ensures that the sum of demands in each cluster $C_j$ is at most equal to the vehicle capacity $C_{\max}$. The second constraint (4) means that each customers $i$ must be served inside its corresponding time window $W_i = [e_i, l_i]$, i.e., $e_i \leq t_i \leq l_i$, $\forall i \in N \cup \{D\}$. Then the depot can be considered as a customer for which the demand is null ($d_D = 0$) and is the only point that is visited twice. The time window $W_D = [e_D, l_D]$ associated to the depot defines the *scheduling horizon* and means that each vehicle cannot leave the depot before its opening (at time $e_D$) and must return to the depot before its closure (at time $l_D$).

## 2 Background

The Vehicle Routing Problem was studied by many authors and several (exact and approximate) methods and strategies were proposed to solve its different versions. For the exact resolution, Azi et al. [3] proposed an exact algorithm, based on column generation and branch-and-price, to solve VRPTW including multiple use of vehicle, i.e., a given vehicle may be associated with several routes. The same authors [2] proposed, several years before, another exact algorithm for a single vehicle routing problem with time windows and multiple routes. Baldacci et al. [4] employed branch-and-price in order to solve a capacitated vehicle routing problem (CVRP) by using an integer programming formulation. New lower bounds were presented and an algorithm to find the optimal solution for CVRP was given. Baldacci and Maniezzo [5] proposed exact methods based on node-routing formulations to tackle the undirected arc-routing problems. Finally, Feillet et al. [11] developed an exact algorithm for the elementary shortest path problem with resource constraints where the authors indicated an application to some vehicle routing problems.

The second category of methods consists to search for approximate solutions by using essentially heuristics and meta-heuristics. Solomon [22] proposed different algorithms in order to solve the vehicle routing and scheduling problems with time window constraints. A two-stage heuristic including ejection pools was for example proposed by Lim and Zhang [16] in order to tackle VRPTW. Chen et al. [9] proposed a heuristic that combines mixed integer programming and a record-to-record travel algorithm in order to solve approximately the *split delivery vehicle routing problem*, a variant of CVRP where a customer may be served by more than one vehicle. Tan et al. [25] proposed several heuristic methods, including simulated annealing, to solve VRPTW.

Insertion heuristics were proposed by Campbell and Savelsbergh [7] for vehicle routing and scheduling problems. Chao et al. [8] proposed a fast heuristic for the *Orienteering Problem* (OP), i.e., a vehicle routing problem where a profit is associated with each customer and the objective is to visit a subset of customers in order to maximize the total benefit and by respecting some constraints. Pisinger and Ropke [19] developed a general heuristic for vehicle routing problems able to solve five different variants of VRP, including CVRP and VRPTW.

Some meta-heuristics are known to be effective for solving the different variants of VRP. For example a genetic algorithm was proposed in [13, 20, 24] for the VRPTW. Ghannadpour et al. [12] studied a dynamic vehicle routing problem with fuzzy time windows (DVRPFTW), i.e., the times windows are not known in advance but arrive dynamically (randomly) over the time. The routing then becomes dynamic. This problem is solved by using a genetic-based algorithm.

Ant colony optimization was adapted for the various variants of VRP. For example Li and Tian [15] tackled the *Open Vehicle Routing Problem* (OVRP), i.e., the vehicles do not return to the depot after serving the last customer. Narasimha et al. [17] used this meta-heuristic to solve the min-max multi-depot vehicle routing problem (MDVRP) where the objective is to minimize the total distance traveled by the vehicles emerging from multiple depots.

Another well-known effective meta-heuristic is Tabu Search that was for example considered by Cordeau et al. [10] for solving VRPTW and by Brandão and Mercer [6] for solving the multi-trip vehicle routing and scheduling problem, i.e., the case where a vehicle may perform several trips. Jiang et al. [14] used tabu search for solving the vehicle routing problem with time windows but with heterogeneous fleet of vehicles.

Finally, Vidal et al. [26] present a survey and synthesis about heuristics developed for the multi-attribute vehicle routing problems. The reader can then refer to this paper for more details about the different methods used for solving VRP and its variants.

In this paper, we propose a three-stage resolution approach-based heuristic for the CVRPTW. The first stage consists to divide the set of customers into $m$ disjoint clusters. The second stage uses beam search in order to compute the shortest path performed by each vehicle inside the corresponding cluster. Each path visits each customer of the given cluster once and only once and the problem constraints must be satisfied, i.e., the time window is not violated and the sum of the demands do not exceed the capacity of the vehicle. In the third and last stage, a local search is applied on the solution obtained at stage 2 in order to try to decrease the total distance traveled by the $m$ vehicles.

# 3 A Three-Stage Heuristic for Solving CVRPTW

It is well known that many methods for solving vehicle routing problems often contain three stages (steps):

**Stage 1:** Divide the set of $n$ customers into $m$ disjoint clusters, i.e., $C_1, ..., C_m$ such that $C_i \cap C_j = \emptyset$ for $1 \leq i < j \leq m$.

**Stage 2:** Construct the shortest feasible path inside each cluster by using a given method. A path is feasible if it satisfies all the constraints.

**Stage 3:** Try to improve the solution obtained after Stage 2 by applying another method such as local search.

## 3.1 The Clustering Stage

This is the first stage in solving our problem (Capacitated Vehicle Routing Problem with Time Windows). This step consists to divide the $n$ customers into $m$ disjoint sets (or clusters). Then each vehicle will visit all the customers of the cluster that is assigned to it. Figure 1 illustrates an example where a set containing 16 points and a depot ($D$) is divided into three disjoint clusters $\{C_1, C_2, C_3\}$.

There exists several methods for clustering and many of them are based on the dispersion (distribution) of the points around a central point called *centroid*. Herein, the well-known *k-means* strategy is chosen in order to compute the clusters. Actually,

**Fig. 1** An example of
clustering a set of points into
three disjoint clusters

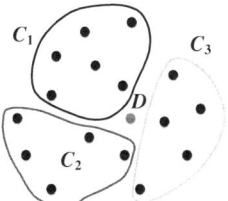

---

**Require:** Set $N$ containing $n$ points (customers);
**Ensure:** $m$ disjoint clusters each of total capacity $\leq C_{max}$;

---

1: *found* $\leftarrow$ false;
2: **while** (*found* = false) **do**
3:  *move* $\leftarrow$ *true*;
4:  Choose randomly $m$ distinct points $\{c_1, ..., c_m\}$ from $N$. Let these points be the $m$ centroids (clusters);
5:  Assign each point $i \in N$ to the nearest centroid $c_j$, $1 \leq j \leq n$;
6:  **while** (*move* = true) **do**
7:   Recompute the coordinates $(x_{c_j}, y_{c_j})$ of each centroid, i.e., each $c_j$ becomes the center of the points assigned to that centroid;
8:   Assign each point $i \in N$ to the nearest centroid $c_j$, $1 \leq j \leq n$;
9:   **if** no point has moved from a cluster to another one **then**
10:    *move* $\leftarrow$ false;
11:   **end if**
12:  **end while**
13:  **if** the capacity of each cluster $\leq C_{max}$ **then**
14:   *found* $\leftarrow$ true;
15:  **end if**
16: **end while**

---

**Algorithm 1** Procedure k-means for CVRPTW

this is an adaptation of k-means in order to compute $m$ clusters each of total capacity smaller than of equal to the capacity $C_{max}$ of the vehicle.

Algorithm 1 explain how procedure k-means works. It receives the set $N$ of customers as input parameter. The procedure's output corresponds to $m$ disjoint clusters respecting the vehicle capacity constraint. k-means begins by choosing $m$ random points $c_1, ..., c_m$ from $N$ (line 4), each point $c_j$, $(1 \leq j \leq m)$ corresponds to a centroid (belonging to cluster $C_j$). After that (at line 5), each of the $|N| - m$ not-assigned yet point is assigned to the nearest centroid $c_j$ of coordinates $(x_{c_j}, y_{c_j})$ in the sense euclidean distance. This means that point $i$ of coordinates $(x_i, y_i)$ is assigned to cluster $C_j$ that minimizes the euclidean distance $\sqrt{(x_i - x_{c_j})^2 + (y_i - y_{c_j})^2}$, for $1 \leq j \leq m$.

After that, in the **while** loop that begins at line 6, the coordinates of the $m$ centroids are recomputed (line 7). This is done by assigning to each $c_j$, $(1 \leq j \leq m)$ the center of the points belonging to cluster $j$. More precisely: $x_{c_j} = \frac{1}{|C_j|} \sum x_{c_k}$ and $y_{c_j} =$

$\frac{1}{|C_j|} \sum y_{c_k}$ for all points $c_k \in C_j$. Each point $i \in N$ is after that assigned to the nearest centroid (among the new computed centroids), this is done in line 8. At line 9, if no point has moved from a centroid to another one, then a stable clustering is obtained: variable *move* is set to the value *false* (line 10) in order to stop recomputing of the centroids. Otherwise, this means that at least one point has moved, then instructions in lines 7–10 are repeated until a stable configuration is obtained. After that, the procedure verifies that the total capacity of each cluster does not exceed the capacity $C_{\max}$ of the vehicles. If so, then variable *found* is set to "true" (line 14) in order to stop the procedure and return $m$ clusters respecting the capacity constraint. If at least one cluster violates the capacity constraint, then the procedure restarts with $m$ other random points by using the first **while** loop (line 2).

Note that the optimal number of clusters ($m$) is not known in the general case. So a dichotomous search can for example be used in order to test several values and determine the best one. Of course, increasing the value of $m$ increases the probability to find clusters respecting the capacity constraint. For well-studied benchmarks, e.g. those proposed by Solomon [22], the same best value of $m$ was found by many authors, so this value can be fixed in advance in order to save computation time.

## 3.2 The Building Stage (Beam Search for Computing Routes)

The second stage takes place once the $m$ clusters were generated by the k-means procedure (Algorithm 1). The objective of the second stage is to compute the shortest path in each cluster. A path corresponds to a route starting at the depot $D$, visiting exactly once each point (customer) are returning after that to the depot. In addition, one vehicle is associated to each cluster. Figure 2 shows an example of solution for the example indicated in Fig. 1.

It is to note that the time windows associated to each customer as well as the capacity of the vehicle make the problem hard to solve, harder than the traveling salesman problem (TSP) in which there are no time windows and no limit to the vehicle capacity.

Remember that the objective in CVRPTW is to minimize the sum of distances traveled by the $m$ vehicles. In order to compute the shortest paths, we propose to use

**Fig. 2** An example of solution obtained after the second stage (beam search)

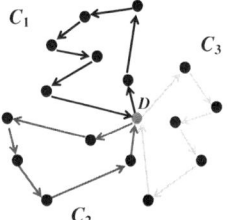

beam search on each cluster. Beam search is a tree search and is a modified version of the well known branch-and-bound method.

Beam search was used to solve different combinatorial problems, such as Scheduling [18] and Cutting-and-Packing problems [1]. In its width—first implementation, the method starts by creating the root node which may contain an initial (starting) partial solution. After that, each node at level $\ell$ generates a set of descendants, these ones correspond to level $\ell + 1$. Each node of the new level is then evaluated by using an evaluation criterion and only a subset of nodes (the $\omega$ best nodes) are retained, the other nodes are discarded. Parameter $\omega$ is known as the *beam width*.

If a node contains a final solution, then this one is evaluated and stored. The corresponding node is after that deleted because no branching is possible from it (leaf). The beam search stops when no branching becomes possible from any node of the current level. The best solution, among the different solutions obtained, is then retained as the final result.

### 3.2.1 Content of a Node in the Search Tree

Let $V = \{v_1, ..., v_{|C_j|}\}$ be the set of vertices (customers) in cluster $C_j$ ($1 \leq j \leq m$).

It is important to define clearly the content of a node in the beam search tree. Each node $\eta_\ell$ at level $\ell$ contains the following elements:

- The set of vertices (customers) already visited $V^+ = \{v_1^+, ..., v_\ell^+\}$.
- The set of vertices that have not yet been visited $V^- = V \backslash V^+$.
- The distance *dist* corresponding to the length of the path $D \rightarrow v_1^+ \rightarrow v_2^+ \rightarrow \cdots \rightarrow v_\ell^+$.

Note then that if a node corresponds to a complete solution, then the path obtained is $D \rightarrow v_1^+ \rightarrow ... \rightarrow v_{|C_j|}^+ \rightarrow D$ and the total distance is the sum of euclidean distances of the corresponding arcs in the path.

As a result, a node $\eta_\ell$ at level $\ell$ in the search tree can be designated by the elements described above, i.e., $\eta_\ell = \{V^+, V^-, dist\}$, where $|V^+| = \ell$ and $|V^-| = |C_j| - \ell$.

### 3.2.2 Selection Criterion for the Next Customer to Visit

As explained above, branching from a node $\eta_\ell$ (or more exactly from the last node $v_\ell^+$ in the path under construction) consists to choose the successors of the vertex $v_\ell^+$ among the vertices in $V^-$. The next vertex $v_i \in V^-$ may be for example the closest one to $v_\ell^+$ in the sense of euclidean distance or the time window interval $[e_i, l_i]$. For example, for the two sets of instances examined in this work (see Sect. 4), the next vertex to visit is the closest one in the sense of parameter $e_i$ (the earliest time) in the time window. For beam search, all the successors $v_i^-$ are ranked in increasing value of parameter $e_i$ and then the $\omega$ first ones are chosen to create $\omega$ distinct branches.

Of course, others criteria were tested, including the latest time $l_i$ and/or the distance between the current customer and the remaining customers to visit, but the experimentations showed that the criterion based on parameter $e_i$ is the best one for the instances tested.

### 3.2.3 Algorithm Beam Search

Algorithm 2 explains how beam search works in order to compute a route. Note that the capacity constraint is not taken into account in the algorithm since the sum of the capacities of the vertices in each cluster is smaller or equal to the vehicle capacity. Indeed, the capacity constraint is always respected after the clustering stage (see Sect. 3.1), then only the time window constraint is checked.

Algorithm 2 receives three input parameters: the cluster $C_j$, i.e., the set of vertices or customers to visit (to serve), the value of the beam width $\omega$, and the selection criterion $\rho$ that will serve to sort the nodes at each level of the tree and then to determine the *best* ones according to this criterion. As output, the algorithm computes the best route (path) $R_j$ of minimal distance beginning at the depot $D$, visiting each customer once, and then returning to the depot.

The root node $\eta_0$ of the search tree is created in line 1. This node contains the set of vertices already visited, i.e., the depot $D$ (so $V^+ = \{D\}$), and the set of vertices not already served $V^- = V$. Since no customer has been already visited, then the distance is equal to 0.

Set $B$ (line 2) corresponds to the nodes at the current level of the search tree. Each node $\eta \in B$ contains a partial route (path) from the depot to a given customer (set $V^+$) as well as the set of remaining customers to visit $V^-$. The total distance to the current customer is also known since this distance is updated each time a new customer is visited and then added to the path (route). $B$ is initialized to $\eta_0$ (line 4). Set $B_{\mathrm{off}}$ (line 3) contains the offspring nodes after branching from each node in $B$.

Since the current level $\ell$ is 0 (root node), then this is indicated in line 5, while the best solution $\eta^*$ is initialized to the root node $\eta_0$ at line 6. The best distance $\eta^*.dist$ is set equal to $+\infty$ (line 7) because this value is to be minimized.

At line 8 the **while** loop starts. So at a given level $\ell$ of the tree, $B$ contains at most $\omega$ distinct partial paths (routes) computed in parallel from the depot (root node). Then branching from a node $\eta_{\ell_i}$ (line 9) consists to explore the successors of the last visited vertex (customer) and to create as many nodes as there are successors with non-violated time windows. So each node in $B$ may have several descendants. Each descendant is then inserted into the set of offspring node $B_{\mathrm{off}}$, that corresponds to level $\ell + 1$. This is why the level is incremented at the next line (10).

After that, at line 11, if there is a node in $B_{\mathrm{off}}$ in which all the customers were served ($V^- = \emptyset$), then the complete solution is computed by adding the returning arc to the depot (line 12). The total distance for the obtained complete solution is then computed and compared to the best known one (line 13). If a lower distance is obtained then the best solution is updated (line 14) and the corresponding node is removed from $B_{\mathrm{off}}$ (line 15).

The most important instruction in beam search is certainly that given in line 18. Indeed, this step consists to sort the nodes according to the selection criterion $\rho$ from

---

**Require:** Cluster $C_j$, the beam width $\omega$, and the selection criterion $\rho$;
**Ensure:** The best shortest route $R_j$ that serves all the customers in cluster $C_1$.

---

1: Let $\eta_0 \leftarrow \{\{D\}, V, 0\}$ be the root node;
2: Let $B$ be the set containing the nodes at a given level of the tree;
3: Let $B_{\text{off}}$ the offspring nodes (descendants of nodes in $B$);
4: $B \leftarrow \{\eta_0\}$;
5: $\ell \leftarrow 0$;
6: $\eta^* \leftarrow \eta_0$; (the best solution found)
7: $\eta^*.dist \leftarrow +\infty$; (best distance)
8: **while** $(B \neq \emptyset)$ **do**
9:   Branch out of each node $\eta_{\ell_i} = \{V_i^+, V_i^-, dist_i\} \in B$ and create the offspring nodes $B_{\text{off}}$ (each node in $B_{\text{off}}$ must respect the time windows);
10:   $\ell \leftarrow \ell + 1$;
11:   **if** $(V_i^- = \emptyset$ for a node $\eta_{\ell_i} \in B_{\text{off}})$ **then**
12:     Add vertex $D$ (depot) to that node and compute the total distance;
13:     **if** $(\eta_{\ell_i}.dist < \eta^*.dist)$ **then**
14:       $\eta^* \leftarrow \eta_{\ell_i}$;
15:       Remove $\eta_{\ell_i}$ from $B_{\text{off}}$;
16:     **end if**
17:   **end if**
18:   Sort the nodes in $B_{\text{off}}$ according to parameter $\rho$ and then keep only the $\min(\omega, |B_{\text{off}}|)$ first nodes, remove the other nodes from $B_{\text{off}}$;
19:   $B \leftarrow B_{\text{off}}$;
20:   $B_{\text{off}} \leftarrow \emptyset$;
21:   **if** there is a node $\eta_{\ell_i} \in B$ for which $V_i^-$ contains a vertex with a violated time window **then**
22:     Remove $\eta_{\ell_i}$ from $B$;
23:   **end if**
24: **end while**

---

**Algorithm 2** Beam Search for computing the shortest path in a cluster

the most important node to the least important one. Then the $\omega$ first nodes are kept and the other ones are removed from $B_{\text{off}}$. Note that if there are less than $\omega$ nodes in $B_{\text{off}}$ then all the nodes are kept. After that set $B_{\text{off}}$ is assigned to $B$ and $B_{\text{off}}$ reset to the empty set (lines 19–20). The last instruction in algorithm 2 consists to remove from $B$ all the nodes that cannot lead to feasible solutions, i.e., that containing violated time windows.

The algorithm stops when set $B$ becomes empty meaning that there is no node to explore or more precisely no customer to serve. Two cases can be distinguished: the algorithm has computed a feasible solution and this one is indicated in node $\eta^*$ as well as the best corresponding distance, or there is no solution (if the distance in node $\eta^*$ is equal to $+\infty$).

Figure 2 shows an example of a solution that may be obtained after the second stage (beam search) on the example (clusters) shown in Fig. 1.

## 3.3 The Local Search Stage for Improving Solution Quality

In order to try to improve the result obtained after the second stage (beam search), a local search is performed on each cluster. This consists to execute the well-known 2-opt algorithm on each cluster (route).

---

**Require:** A route $R = v_0 \rightarrow v_1 \rightarrow \ldots \rightarrow v_{|C_j|} \rightarrow v_{|C_j|+1}$;
**Ensure:** A route $R'$ with a length at most equal to that of $V$;

---

1: *improvement* ← true;
2: **while** (*improvement* = true) **do**
3:   *improvement* ← false;
4:   **for each** vertex $v_i \in R$ **do**
5:     **for each** vertex $v_j \in R$ ($j \neq i - 1, j \neq i + 1$) **do**
6:       **if** (dist($v_i, v_{i+1}$) + dist($v_j, v_{j+1}$) > dist ($v_i, v_j$) + dist($v_{i+1}, v_{j+1}$) **AND** the time
         windows will not be violated) **then**
7:         Replace arcs ($v_i \rightarrow v_{i+1}$) and ($v_j \rightarrow v_{j+1}$)
             by arcs ($v_i \rightarrow v_j$) and ($v_{i+1} \rightarrow v_{j+1}$);
8:         *improvement* ← true;
9:       **end if**
10:     **end for**
11:   **end for**
12: **end while**

---

**Algorithm 3** 2-opt algorithm

2-opt is an iterative method that consists, at each iteration, to *break* two nonconsecutive arcs in the route and to link the four extremities in order to form another path and by respecting the time windows of course. The replacement is kept if the obtained solution is better.

The 2-opt method is given in Algorithm 3. In each iteration, the algorithm examines each two distinct and non-adjacent arcs $v_i \rightarrow v_{i+1}$ and $v_j \rightarrow v_{j+1}$ in the route $R$. These two arcs are replaced by the arcs $v_i \rightarrow v_j$ and $v_{i+1} \rightarrow v_{j+1}$ if and only if the distance decreases and the time windows are not violated. This process is repeated as long as there is improvement. Figure 3 shows an example of improvement obtained by the 2-opt procedure on the solution of Fig. 2. The arcs that had changed are indicated in dotted lines.

**Fig. 3** A solution obtained after the third stage (local search)

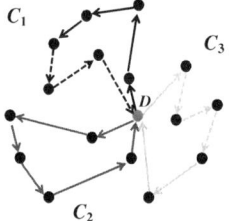

## 3.4 The Three-Stage Algorithm (3PA) for Solving CVRPTW

The three-stage algorithm based on clustering, beam search, and 2-opt local search is given in algorithm 4. It receives as input parameters the set of customers $N$, the depot $D$, and the number of vehicles. The algorithm's output corresponds to a set containing $m$ feasible routes (respecting the constraints) of minimum distance, each one starts and ends at the depot $D$.

At line 1, the clustering stage (Algorithm 1) is called in order to create $m$ distinct clusters. The selection criterion ($\rho$), that serves to choose the next customer to serve is set at line 2. Then, Algorithm 2 (beam search) is executed on each cluster (line 5), and this for several values of the beam width, i.e., for all values $\omega \in [1, ..., \omega_{max}]$. The local search (Algorithm 3) is then executed (line 6) on each solution computed by beam search.

---

**Require:** A set $N = \{1, ..., n\}$ of customers, the depot $D$, and $m$ the number of vehicles (clusters).
**Ensure:** A set of routes minimizing the total distance and respecting the capacity and the time windows constraints.

---

1: Call the clustering stage (Algorithm 1) and create $m$ clusters $\{C_1, ..., C_m\}$ respecting the vehicle capacity constraint;
2: Define the selection criterion $\rho$;
3: **for each** cluster $C_j$, $(1 \leq j \leq m)$ **do**
4:   **for** $\omega = 1$ to $\omega_{max}$ **do**
5:     Call Algorithm 2: Beam-Search($C_j, \omega, \rho$);
6:     Apply Algorithm 3 (2-opt) on the solution returned by Beam-Search;
7:   **end for**
8: **end for**

---

**Algorithm 4** The three-stage algorithm 3PA for solving CVRPTW

## 4 Computational Results

The proposed method is coded with the C++ language and the program run under Microsoft Windows environment on a computer with 2 GB of RAM and a 2.26 GHz Intel processor.

Two sets of instances were tested, namely C1 and C2, proposed by Solomon [23]. Table 1 indicates the characteristics of these instances. Each instance of each set contains 100 customers (column 2), they have also all the same service time (time needed to serve a customer) which is equal to 90 (column 4). The depot $D$ has also the same coordinates for all the instances.

The first common characteristic for all the instances of the same set is the customer *demand*, i.e., the quantity to deliver to each customer. This value is fixed in each set for a given customer. More precisely, for two distinct instances in the same set (C1 or C2) each customer $i$ has the same demand $d_i$. The second common characteristic

**Table 1** Characteristics of the C1 and C2 instances

| Set | Number of customers | Scheduling horizon | Service time | Vehicle capacity |
|-----|---------------------|--------------------|--------------|------------------|
| C1  | 100                 | 1236               | 90           | 200              |
| C2  | 100                 | 3390               | 90           | 700              |

is that a given customer $i$ has the same coordinates $(x_i, y_i)$ in two distinct instances of the same set (C1 or C2).

The third common characteristic appears in the scheduling horizon (column 3) of Table 1, which is *short* for instances of set C1 (1,236) and *large* for the instances of set C2 (3,390). This means for example that the tours in instances C1 will all finish at most after 1,236 units of time and at most after 3,390 units of time for the instances of set C2. Finally, the fourth common characteristic concerns the vehicle capacity (column 5) of Table 1. In set C1, vehicles of capacity $C_{max} = 200$ are used while this capacity is equal to 700 for instances of set C2.

From these characteristics Solomon designed several instances in the same set by changing the time windows from an instance to another one. More precisely, there are nine instances C101–C109 in the first set C1 and eight instances C201–C208 in the second set C2. Two distinct instances in the same set (C1 or C2) have:

- the same coordinates for a given customer $i$ as well as for the depot $D$
- the same demand for a given customer $i$
- the same vehicle capacity
- different time windows

Figure 4 shows the distribution of the time windows for the 100 customers of instance C206. Each time window is represented by a vertical bar indicating the

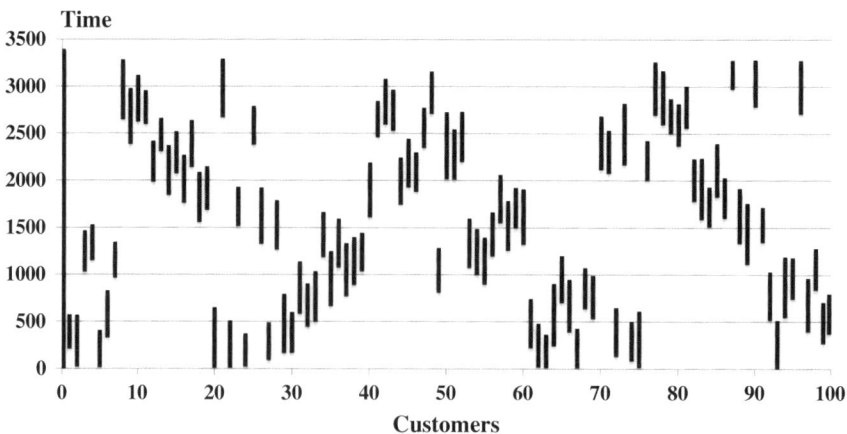

**Fig. 4** Time windows associated with the 100 customers of instance C206 (scheduling horizon = 3390)

earliest time and the latest time that correspond to the ready time and due date respectively. Note that the first bar (customer 0) corresponds to the depot, the corresponding time window begins at 0 and ends at 3,390. These values determine the scheduling horizon for the instance.

For more details, the reader can refer to Solomon's web site [23].

Then, one can surmise that less vehicles (clusters) will be needed for instances of set C2 comparing to set C1 because of the larger value of the capacity and the greater length of the time windows. This is in fact the case as proven in different works published in the literature.

Finally, note that the value of $m$ (number of clusters or vehicles) is fixed to the best value found by several authors in the literature. More precisely, $m = 10$ for instances C1 and $m = 3$ for set C2.

Table 2 indicates the results obtained on the two sets of instances (C1 and C2) and shows the importance of the third stage of algorithm 3PA, i.e., the 2-opt strategy. Column 1 in Table 2 gives the name of each instance. Column 2 gives the distance obtained after the second step of algorithm 3PA, i.e., the Beam Search algorithm (namely BS) and column 3 indicates the value of the distance obtained after the third stage (2-opt), the improved values are displayed in bold. Finally, the last column in Table 2 shows the percentage of improvement after the 2-opt stage. Let for example $Dist_{BS}$ be the distance obtained by the beam search (BS) and

**Table 2** Comparison between the results obtained after the second step of algorithm 3PA (Beam Search) and those obtained after the third stage (2-opt) on instances C1 and C2

| Instance | Beam Search (BS) | BS +2-opt | Imp. (%) |
|---|---|---|---|
| C101 | 828.94 | 828.94 | 0.00 |
| C102 | 878.47 | **828.94** | 5.64 |
| C103 | 862.53 | **828.94** | 3.89 |
| C104 | 866.87 | **828.94** | 4.38 |
| C105 | 828.94 | 828.94 | 0.00 |
| C106 | 828.94 | 828.94 | 0.00 |
| C107 | 828.94 | 828.94 | 0.00 |
| C108 | 832.9 | **828.94** | 0.48 |
| C109 | 828.94 | 828.94 | 0.00 |
| C201 | 591.56 | 591.56 | 0.00 |
| C202 | 697.22 | **591.56** | 15.15 |
| C203 | 644.43 | **591.17** | 8.26 |
| C204 | 647.12 | **591.17** | 8.65 |
| C205 | 588.88 | **588.49** | 0.07 |
| C206 | 597.35 | **588.49** | 1.48 |
| C207 | 650.66 | **588.32** | 9.58 |
| C208 | 588.88 | 588.88 | 0.00 |
| Average | | | 3.39 |

$\text{Dist}_{BS+2-opt}$ be the distance obtained after 2-opt, then the improvement is computed as $\frac{\text{Dist}_{BS}-\text{Dist}_{BS+2-opt}}{\text{Dist}_{BS}} \times 100\,\%$.

From Table 2 we can see that 2-opt has improved BS for ten solutions out of seventeen (58.8 %). More precisely, four solutions were improved in set C1 and six solutions were improved in set C2. The average percentage of improvement is equal to 3.39 %.

Table 3 shows the results obtained on the 17 instances where column 1 indicates the name of each instance. Columns 2–4 contain the best known results in the literature (to our knowledge). Column 2 shows the best value for $m$ (the number of vehicles) and column 3 the best distance. Column 4 (Ref.) indicates the reference to the paper where the best values for $m$ and Dist were obtained. Columns 5–7 contain the results obtained by a method based on goal programming and genetic algorithm (GP-GA) proposed by Ghoseiri and Ghannadpour [13]. So columns 5 and 6 indicate the best value for $m$ and the best distance respectively. Column 7 corresponds to the gap between the distance obtained by GP-GA (column 6) and the best known distance in the literature (column 3). This gap is computed as follows: gap $= 100\,\% \times (\text{Dist}_{bestknown} - \text{Dist}_{GP-GA})/\text{Dist}_{bestknown}$. Columns 8–12 summarize the results obtained by the proposed algorithm 3PA. Column 8 indicates the minimum number of vehicles $m$ while column 9 shows the best minimum distance obtained by algorithm 3PA. Column 10 ($\omega_{max}$) corresponds to the maximum value

**Table 3** Results obtained on instances C1 and C2

| Inst. | Best known | | | GPA-GA [13] | | | The proposed method (3PA) | | | | |
|---|---|---|---|---|---|---|---|---|---|---|---|
| | $m$ | Dist. | Ref. | $m$ | Dist. | Gap (%) | $m$ | Dist. | $\omega_{max}$ | Time (s) | Gap (%) |
| C101 | 10 | **828.94** | [13] | 10 | **828.94** | 0.0 | 10 | **828.94** | 50 | 29 | 0.00 |
| C102 | 10 | **828.94** | [13] | 10 | **828.94** | 0.0 | 10 | **828.94** | 50 | 66 | 0.00 |
| C103 | 10 | **828.06** | [13] | 10 | **828.06** | 0.0 | 10 | 828.94 | 50 | 140 | −0.11 |
| C104 | 10 | **824.78** | [13] | 10 | **824.78** | 0.0 | 10 | 828.94 | 50 | 183 | −0.50 |
| C105 | 10 | **828.94** | [13] | 10 | **828.94** | 0.0 | 10 | **828.94** | 50 | 36 | 0.00 |
| C106 | 10 | **828.94** | [13] | 10 | **828.94** | 0.0 | 10 | **828.94** | 50 | 41 | 0.00 |
| C107 | 10 | **828.94** | [13] | 10 | **828.94** | 0.0 | 10 | **828.94** | 50 | 41 | 0.00 |
| C108 | 10 | **828.94** | [13] | 10 | **828.94** | 0.0 | 10 | **828.94** | 1,600 | 146,600 | 0.00 |
| C109 | 10 | **828.94** | [13] | 10 | **828.94** | 0.0 | 10 | **828.94** | 50 | 80 | 0.00 |
| C201 | 3 | **591.56** | [13] | 3 | **591.56** | 0.0 | 3 | **591.56** | 50 | 176 | 0.00 |
| C202 | 3 | **591.56** | [13] | 3 | **591.56** | 0.0 | 3 | **591.56** | 50 | 240 | 0.00 |
| C203 | 3 | **591.17** | [13] | 3 | **591.17** | 0.0 | 3 | **591.17** | 100 | 13,210 | 0.00 |
| C204 | 3 | **590.60** | [20] | 3 | 599.96 | −1.58 | 3 | 591.17 | 100 | 12,390 | −0.10 |
| C205 | 3 | **588.16** | [24] | 3 | 588.88 | −0.12 | 3 | 588.49 | 50 | 528 | −0.06 |
| C206 | 3 | **588.49** | [20] | 3 | 588.88 | −0.07 | 3 | **588.49** | 2,000 | 286, 700 | 0.00 |
| C207 | 3 | **588.29** | [21] | 3 | 591.56 | −0.56 | 3 | 588.32 | 2,000 | 291,600 | −0.01 |
| C208 | 3 | **588.32** | [21] | 3 | **588.32** | 0.0 | 3 | 588.88 | 50 | 618 | −0.09 |
| Av. dev. | | | | | | −0.14 | | | | | −0.05 |

of the beam width used for each instance, then $\omega_{max} = 50$ means that beam search was executed for each value $1 \leq \omega \leq 50$. The next column (time) indicates the total computation time needed for the execution of algorithm 3PA (in seconds). The last column (gap) indicates the difference (in %) between the solution obtained by the proposed method 3PA and the best known solution in the literature (column 3). More precisely gap $= 100\% \times (\text{Dist}_{\text{Best known}} - \text{Dist}_{\text{3PA}})/\text{Dist}_{\text{best known}}$. Finally, the last row of Table 3 indicates the average gap for the two compared methods (GP-GA and 3PA). As expected, the number of vehicles needed for C1 instances (10) is larger than that needed for instances of set C2 (only 3 vehicles).

The results of Table 3 indicate that the proposed method 3PA reached the best known results in 11 cases out of 17. In the six other cases, the result is very close to the best known value since the gap is often smaller or equal to $-0.11\%$, except for instance C104 where the gap reaches $-0.50\%$. Note that even if the GP-GA method reaches the best known results in 13 cases out of 17, its average gap $(-0.14\%)$ is worst then that obtained by algorithm 3PA $(-0.05\%)$.

Concerning the computation time of algorithm 3PA, it is at most 183 seconds for instances of set C1 (except for C108 which was hard to solve). Each cluster in instances of set C1 contains about 10 customers. The computation time is generally greater for the second set C2, this is due to larger number of customers in each cluster (which is about 30) and then the number of combinations in each cluster (route) becomes larger.

But how to determine the maximum value for the beam width $\omega_{max}$, especially for new instances. One can for example fix the maximum value to 50 or 100 or use a limited computation time.

Table 4 indicates, for each of the nine solutions of instances C1 indicated in Table 3, the best value $\omega^*$ that gave the best solution for each cluster $C_j$, $j = 1, .., m$. We can see for example that $w^* = 1$ for all $j = 1, .., 10$ for instance C101. Table 5 gives the same information than Table 4 but for the eight instances of group C2 (three clusters).

**Table 4** Best value for the beam width ($\omega^*$) in each cluster for instances of group C1

| Inst. | Cluster | | | | | | | | | |
|-------|---|---|---|---|---|---|---|---|---|----|
|       | 1 | 2 | 3 | 4 | 5 | 6 | 7 | 8 | 9 | 10 |
| C101  | 1 | 1 | 1 | 1 | 1 | 1 | 1 | 1 | 1 | 1 |
| C102  | 7 | 7 | 1 | 1 | 1 | 6 | 7 | 6 | 8 | 1 |
| C103  | 39 | 13 | 1 | 8 | 7 | 10 | 44 | 5 | 8 | 4 |
| C104  | 24 | 5 | 1 | 5 | 7 | 8 | 14 | 5 | 8 | 7 |
| C105  | 1 | 1 | 1 | 1 | 1 | 1 | 1 | 1 | 1 | 1 |
| C106  | 8 | 1 | 1 | 1 | 1 | 1 | 1 | 1 | 1 | 2 |
| C107  | 1 | 1 | 1 | 1 | 1 | 1 | 1 | 1 | 1 | 1 |
| C108  | 1518 | 2 | 3 | ? | ? | 10 | 1 | 1 | 3 | 1 |
| C109  | 1 | 1 | 1 | 2 | 1 | 1 | 1 | 1 | 1 | 2 |

**Table 5** Best value for the beam width ($\omega^*$) in each cluster for instances of group C2

|        |       | Cluster |       |
| Inst.  | 1     | 2       | 3     |
|--------|-------|---------|-------|
| C201   | 1     | 1       | 1     |
| C202   | 43    | 19      | 26    |
| C203   | 43    | 35      | 68    |
| C204   | 43    | 22      | 68    |
| C205   | 1     | 1       | 1     |
| C206   | 6     | 1,687   | 1     |
| C207   | 43    | 1,928   | 1     |
| C208   | 1     | 1       | 1     |

Note that the values of $\omega^*$ are heterogeneous for some instances. This is the case for example for instances C103 and C104. This means that these two instances are harder to solve (due essentially to the characteristics of the time windows). In addition, some instances are very easy to solve, these ones are those for which the values of $\omega^*$ are all equal (or close) to 1. This is the case for example for instances C101, C105, C107, C109, C205, and C208.

Figure 5 shows an example of solution obtained on instance C206 after the second step of the algorithm (beam search), i.e., the output of Algorithm 2. This corresponds to Fig. 5a. The total distance obtained after this step is equal to 597.35. Figure 5b indicates improvement of the solution of Fig. 5a by the 2-opt procedure (Algorithm 3). The new distance was decreased from 597.35 to 588.49, i.e., an improvement of

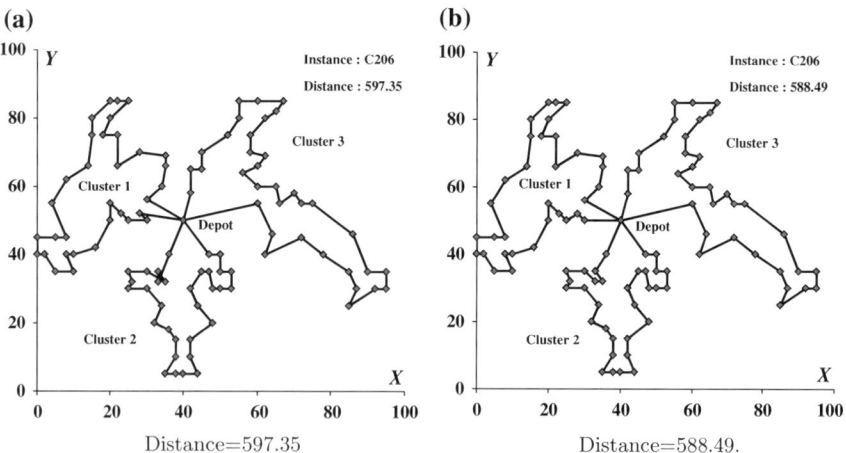

**Fig. 5** Solution obtained by algorithm 3PA on instance C206. **a** After the second step (Beam Search) and **b** after the third step (2-Opt)

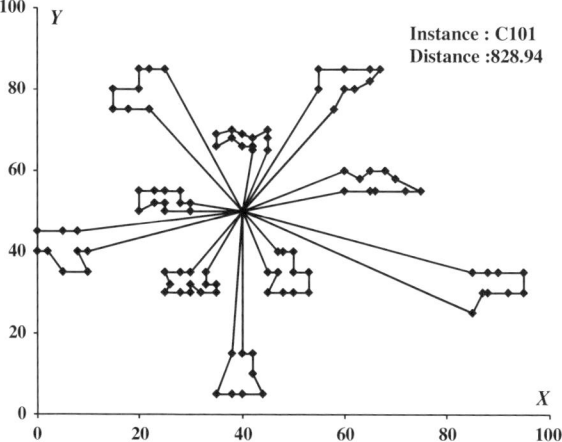

**Fig. 6** Solution obtained by algorithm 3PA on instance C101 after the third step (Local Search): $m = 10$, Distance $= 828.94$

1.48 %. The 2-opt stage has changed several arcs in clusters 1 and 2 while the third cluster has not changed.

Figure 6 displays the solution obtained after the third stage of the proposed algorithm on instance C101. There are ten clusters and the total distance computed is equal to 828.94. This value coincides to the best known solution in the literature.

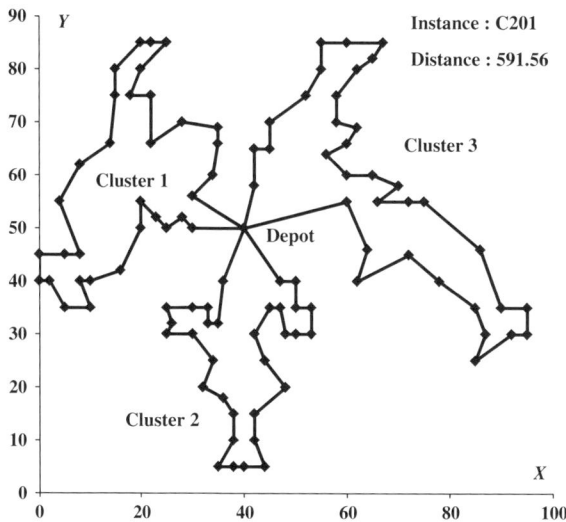

**Fig 7** Solution obtained by algorithm 3PA on instance C201 after the third step (Local Search): $m = 3$, Distance $= 591.56$

Finally, Fig. 7 shows the final solution obtained by algorithm 3PA on instance C201. Note that even if the coordinates of the customers are the same than in C206 (Fig. 5b), some arcs in the two solutions differ. These arcs are situated in clusters 2 and 3. Cluster 1 is the same in the two solutions.

## 5 Conclusion and Future Work

In this paper an algorithm, denoted by 3PA, is proposed in order to solve the capacitated vehicle routing problem with time windows (CVRPTW). 3PA contains three steps. After creating disjunctive clusters in the first stage by using the well-known k-means method, beam search is executed in order to compute the shortest path inside each cluster. The most characteristic of beam search is that it explores several paths in parallel and increases then the probability to find *good* paths. The last stage is a local search that executes the 2-opt strategy in order to try to improve the solution given by beam search. The results obtained on the instances used show that the method is competitive since the computations revealed that algorithm 3PA was better than a method based on goal programming and genetic algorithm (GP-GA). Indeed, 3PA obtained a gap closer to the best known results in the literature than the gap obtained by GP-GA.

As a future work, it will be interesting to add a global evaluation criterion for beam search in order to provide solutions of better quality before calling the local search (third) stage and then to improve the overall solution. It will be also interesting to improve the local search by interchanging points (customers) between clusters.

## References

1. Akeb, H., Hifi, M.: Algorithms for the circular two-dimensional open dimension problem. Int. Trans. Oper. Res. **15**, 685–704 (2008)
2. Azi, N., Gendreau, M., Potvin, J.-Y.: An exact algorithm for a single vehicle routing problem with time windows and multiple routes. Eur. J. Oper. Res. **178**, 755–766 (2007)
3. Azi, N., Gendreau, M., Potvin, J.-Y.: An exact algorithm for a vehicle routing problem with time windows and multiple use of vehicles. Eur. J. Oper. Res. **202**, 756–763 (2010)
4. Baldacci, R., Hadjiconstantinou, E.A., Mingozzi, E.A.: An exact algorithm for the capacitated vehicle routing problem based on a two-commodity network flow formulation. Oper. Res. **52**, 723–738 (2004)
5. Baldacci, R., Maniezzo, V.: Exact methods based on node-routing formulations for undirected arc-routing problems. Networks. **47**, 52–60 (2006)
6. Brandão, J.C.S., Mercer, A.: A tabu search algorithm for the multi-trip vehicle routing and scheduling problem. Eur. J. Oper. Res. **100**, 180–191 (1997)
7. Campbell, A.M., Savelsbergh, M.: Efficient insertion heuristics for vehicle routing and scheduling problems. Transp. Sci. **38**, 369–378 (2004)
8. Chao, I.M., Golden, B.L., Wasil, E.A.: A fast and effective heuristic for the orienteering problem. Eur. J. Oper. Res. **88**, 101–111 (1996)

9. Chen, S., Golden, B., Wasil, E.: The split delivery vehicle routing problem: applications, algorithms, test problems, and computational results. Networks **49**, 661–673 (2007)
10. Cordeau, J.-F., Laporte, G., Mercier, A.: A unified tabu search heuristic for vehicle routing problems with time windows. J. Oper. Res. Soc. **52**, 928–936 (2001)
11. Feillet, D., Dejax, P., Gendreau, M., Gueguen, C.: An exact algorithm for the elementary shortest path problem with resource constraints: application to some vehicle routing problems. Networks **44**, 216–229 (2004)
12. Ghannadpour, S.F., Noori, S., Tavakkoli-Moghaddam, R., Ghoseiri, K.: A multi-objective dynamic vehicle routing problem with fuzzy time windows: model, solution and application. Appl. Soft Comput. **14**(C), 504–527 (2014)
13. Ghoseiri, K., Ghannadpour, S.F.: Multi-objective vehicle routing problem with time windows using goal programming and genetic algorithm. Appl. Soft Comput. **10**(4), 1096–1107 (2010)
14. Jiang, J., Ng, K.M., Poh, K.L., Teo, K.M.: Vehicle routing problem with a heterogeneous fleet and time windows. Expert Syst. Appl. **41**(8), 3748–3760 (2014)
15. Li, X., Tian, P.: An ant colony system for the open vehicle routing problem. Lecture Notes in Computer Science vol. 4150 (Springer, Heidelberg, 2006), pp. 356–363
16. Lim, A., Zhang, X.: A two-stage heuristic with ejection pools and generalized ejection chains for the vehicle routing problem with time windows. J. Comput. Inform. **19**(3), 443–457 (2007)
17. Narasimha, K.V., Kivelevitch, E., Sharma, B., Kumar, M.: An ant colony optimization technique for solving minmax Multi-Depot Vehicle Routing Problem. Swarm Evol. Comput. **13**, 63–73 (2013)
18. Ow, P.S., Morton, T.E.: Filtered beam search in scheduling. Int. J. Prod. Res. **26**(1), 35–62 (1988)
19. Pisinger, D., Ropke, S.: A general heuristic for vehicle routing problems. Comput. Oper. Res. **34**, 2403–2435 (2007)
20. Potvin, J.Y., Bengio, S.: The vehicle routing problem with time windows. Part II. Genetic search. J. Comput. Inform. **8**, 165–172 (1996)
21. Rochat, Y., Taillard, E.D.: Probabilistic diversification and intensification in local search for vehicle routing. J. Heurisctics. **1**, 147–167 (1995)
22. Solomon, M.M.: Algorithms for the vehicle routing and scheduling problems with time window constraints. Oper. Res. **35**(2), 254–265 (1987)
23. Solomon, M.M.: VRPTW benchmark problems. http://w.cba.neu.edu/~msolomon
24. Tan, K.C., Chew, Y.H., Lee, L.H.: A hybrid multiobjective evolutionary algorithm for solving vehicle routing problem with time windows. Comput. Optim. Appl. **34**, 115–151 (2006)
25. Tan, K.C., Lee, L.H., Zhu, Q.L., Ou, K.: Heuristic methods for vehicle routing problem with time windows. Artif. Intell. Eng. **15**, 281–295 (2001)
26. Vidal, T., Crainic, T.G., Gendreau, M., Prins, C.: Heuristics for multi-attribute vehicle routing problems: a survey and synthesis. Eur. J. Oper. Res. **231**(1), 1–21 (2013)

# Fuzzy Bicriteria Optimization Approach to Distribution Network Design

Santiago García-Carbajal, Antonio Palacio, Belarmino Adenso-Díaz
and Sebastián Lozano

**Abstract** Distribution network design deals with defining which elements will be part of the supply chain and how they will be interrelated. Many authors have studied this problem from a cost minimization point of view. Nowadays the sustainability factor is increasing its importance in the logistics operations and must be considered in the design process. We deal here with the problem of determining the location of the links in a supply chain and the assignment of the final customers considering at the same time cost and environmental objectives. We use a fuzzy bicriteria model for solving the problem, embedded in a genetic algorithm that looks for the best trade-off solution. A set of experiments have been carried out to check the performance of the procedure, using some instances for which we know a priori a good reference solution.

## 1 Introduction

The fierce competition between the different supply chains makes it necessary that efficiency be continuously pursued. One of the most important strategic decisions, and one that has a long-term impact in the economic results of the logistics operations, is the design of the distribution network.

Distribution network design is the process of determining the structure of a supply chain, defining which elements will be part of it (i.e., where locate the facilities), and what will be the interrelationships between them (i.e., the allocation of customers to facilities and how the material and products will flow in the network between the nodes in the network). For that reason the problem is often called location-allocation (e.g. [1]).

S. García-Carbajal · A. Palacio · B. Adenso-Díaz
Escuela Politécnica Superior de Ingeniería de Gijón, Universidad de Oviedo,
33003 Oviedo, Asturias, Spain

S. Lozano (✉)
Escuela Superior de Ingenieros, Universidad de Sevilla, 41092 Seville, Spain
e-mail: slozano@us.es

© Springer International Publishing Switzerland 2015
S. Fidanova (ed.), *Recent Advances in Computational Optimization*,
Studies in Computational Intelligence 580, DOI 10.1007/978-3-319-12631-9_2

**Structure of warehouses**

| | | Decentralised | Centralised |
|---|---|---|---|
| **Size of the company** | **Large** | - Cost of lost sales | - Balance peaks of demand |
| | | - Delivery time | - Delivery precision |
| | | - Local exposure | - Inventory level |
| | | - Service level | - Number of employees |
| | | - Transportation Costs | - Warehousing costs |

**Fig. 1** Major drivers for centralization versus decentralization in logistic networks design, according to [2]

The decision of choosing a centralized supply network or a distributed one, is a classical problem in logistics management [2]. Many works [3, 4] have discussed the advantages of each alternative, from the point of view of total costs. Usually, centralizing means (see Fig. 1) a better use of resources (inventory and personnel), lower costs and a better management of demand fluctuations, while decentralization implies a better service, better fill rate and reduced transportation costs.

Considering cost as the main (and only) design criterion in logistic networks, is the classical approach used by most firms. However, nowadays some other issues (such as the robustness of the network, or the environmental impacts caused by the logistics operations) are also important factors to be considered. New environmental regulations issued by governments and the need to improve the quality of customer service level have led companies to seek more sustainable supply chains, and supply chains with lower risks of causing delays in delivery of products to customers. In this way, research works that combine these objectives with cost minimization are becoming more frequent [5–7].

However more approaches for the problem have been defined beyond the centralization/decentralization issue. In [8], Akkerman et al. consider, from a hierarchical point of view, a second level called distribution network planning that includes the decisions related to fulfilling the aggregate demand (i.e., aggregate product flows and delivery frequencies).

Many authors have studied these problems, most of them (around two thirds according to [9]) by considering as the single objective function the minimization of the costs involved in the process. However, as mentioned above, for different reasons (legal pressure, customers demand, ethical consciousness, etc.) nowadays the sustainability factor is increasing its importance in the business management and specifically in the logistics operations, where transportation of goods is a high pollutant activity. It is therefore necessary to consider the operations impact when defining the distribution network.

The aim of this study is the formulation of a model and a solution procedure for the location-allocation problem when two criteria (cost and environmental impact of transportation) are considered at the same time.

## 2 Problem Setting

Let us suppose that there is an uncapacitated central plant that must distribute a single product among many customers. Those customers have uncertain (i.e. fuzzy) demands. We need to define the distribution network, choosing the capacitated intermediate warehouses to set up, and allocating each customer to one of warehouses (or to the central facility). There are two types of vehicles: large trucks (used for high demand customers and for serving the warehouses from the central plant) and smaller trucks (i.e. vans).

We are going to consider two objective functions. One is the minimization of the logistics costs (transportation and warehouses set-up). The transportation costs will be proportional to distances and depend on type of truck used. The second objective function is the minimization of the environmental impact of the Greenhouse Gases (GHG) emissions (e.g. $CO_2$) due to transportation.

Note that, in principle, every customer could be served from the central facility, but if the demand is small, the cost and environmental impact of such direct shipments would be very high, likely bigger that delivering the goods from a near warehouse.

Our problem consists in deciding which of the potential warehouse locations will be opened and from which warehouse should each customer be served (considering the limited capacities of the warehouses) in such a way that total cost and GHG emissions are minimized.

## 3 Model Formulation

Table 1 shows the notation used for modelling the problem. Note that the set of potential warehouse locations are given together with the distance, unit transport cost and unit GHG emissions factor from the central plant to each potential warehouse location $j$. From each potential warehouse location $j$ only a subset of customers $I(j)$ can be served. The distance, unit transport cost and unit GHG emissions factors from each warehouse location to each customer $i \in I(j)$ are given.

The membership function of the demand of each customer $i$ is given by a Triangular Fuzzy Number (TFN) with parameters $\left(D_i^-, D_i^0, D_i^+\right)$. Each warehouse has a capacity, i.e. an upper bound on the flow of goods that it can convey from the central plant to its allocated customers. The membership function of the capacity of warehouse $j$ is given by a linear decreasing function with parameters $\left(U_j^-, U_j^+\right)$. Each warehouse also has a lower bound on the flow that it should handle in case it is selected. This minimum flow is imposed to guarantee an economic operation of the warehouse. The membership function of the minimum flow of warehouse $j$ is given by a linear increasing function with parameters $\left(L_j^-, L_j^+\right)$.

**Table 1** Notation

| Notation | Description |
|---|---|
| $i$ | Index of customer ($i = 1, \ldots, N$) |
| $j$ | Index on potential warehouse locations ($j = 1, \ldots, A$) |
| $I(j), \hat{I}$ | Subsets of customers that can be served from warehouse $j$ and from the central plant, respectively |
| $\tilde{D}_i$ | Fuzzy demand of customer $i$. A Triangular Fuzzy Number membership function is assumed |
| | $$\mu_{D_i}(x) = \begin{cases} 0 & \text{if } x \leq D_i^- \\ \dfrac{x - D_i^-}{D_i^0 - D_i^+} & \text{if } D_i^- \leq x \leq D_i^0 \\ \dfrac{D_i^+ - x}{D_i^+ - D_i^0} & \text{if } D_i^0 \leq x \leq D_i^+ \\ 0 & \text{if } x \geq D_i^+ \end{cases}$$ |
| $\tilde{U}_j$ | Fuzzy capacity of warehouse $j$. A decreasing linear membership function is assumed. |
| | $$\mu_{U_j}(x) = \begin{cases} 1 & \text{if } x \leq U_j^- \\ \dfrac{U_j^+ - x}{U_j^+ - U_j^-} & \text{if } U_j^- \leq x \leq U_j^+ \\ 0 & \text{if } x \geq U_j^+ \end{cases}$$ |
| $\tilde{L}_j$ | Fuzzy minimum flow of warehouse $j$. An increasing linear membership function is assumed (with parameter $L_j^+ \ll U_j^-$). |
| | $$\mu_{L_j}(x) = \begin{cases} 0 & \text{if } x \leq L_j^- \\ \dfrac{x - L_j^-}{L_j^+ - L_j^-} & \text{if } L_j^- \leq x \leq L_j^+ \\ 1 & \text{if } x \geq L_j^+ \end{cases}$$ |
| $f_j$ | Fixed cost of warehouse $j$ |
| $c_{ji}$ | Unit transport cost between warehouse $j$ and customer $i$ |
| $\hat{c}_i$ | Unit transport cost between central plant and customer $i$ |
| $\hat{\hat{c}}_j$ | Unit transport cost between central plant and warehouse $j$ |
| $e_{ji}$ | Unit GHG emissions factor for transport between warehouse $j$ and customer $i$ |
| $\hat{e}_i$ | Unit GHG emissions factor for transport between central plant and customer $i$ |
| $\hat{\hat{e}}_j$ | Unit GHG emissions factor for transport between central plant and warehouse $j$ |
| $t_{ji}$ | Distance between warehouse $j$ and customer $i$ |
| $\hat{t}_i$ | Distance between central plant and customer $i$ |
| $\hat{\hat{t}}_j$ | Distance between central plant and warehouse $j$ |
| $x_{ji}$ | Amount of product shipped from warehouse $j$ to customer $i \in I(j)$ |
| $\hat{x}_i$ | Amount of product shipped from central plant to customer $i$ |
| $y_j$ | Amount of product shipped from central plant to warehouse $j$ |

The proposed bicriteria optimization model consist in the minimization of both cost and GHG emissions:

$$\min \sum_j f_j y_j + \sum_{i \in \hat{I}} \hat{c}_i \hat{t}_i \hat{x}_i + \sum_j \sum_{i \in I(j)} \left( \hat{\hat{c}}_j \hat{t}_j + c_{ji} t_{ji} \right) x_{ji} \tag{1}$$

$$\min \sum_{i \in \hat{I}} \hat{e}_i \hat{t}_i \hat{x}_i + \sum_j \sum_{i \in I(j)} \left( \hat{\hat{e}}_j \hat{t}_j + e_{ji} t_{ji} \right) x_{ji} \tag{2}$$

subject to

$$\hat{x}_i + \sum_{j:i \in I(j)} x_{ji} = \tilde{D}_i \qquad \forall i \in \hat{I} \tag{3}$$

$$\sum_{j:i \in I(j)} x_{ji} = \tilde{D}_i \qquad \forall i \notin \hat{I} \tag{3'}$$

$$\tilde{L}_j y_j \leq \sum_{i \in I(j)} x_{ji} \leq \tilde{U}_j y_j \tag{4}$$

$$\hat{x}_i \geq 0 \ \forall i \in \hat{I} \quad x_{ji} \geq 0 \ \forall j, i \in I(j) \quad y_j \in \{0, 1\} \ \forall j \tag{5}$$

In order to solve this model, a Fuzzy Multiobjective Optimization approach based on the additive model of Tiwari [10] is proposed. Thus, the new objective function, to be maximized, will be the sum of the membership functions of the fuzzy constraints and of the two objective functions. The latter are fuzzified using decreasing linear membership functions, between the thresholds $(C^-, C^+)$ and $(E^-, E^+)$, respectively (see Fig. 2). These total cost and total emissions thresholds are evaluated in the following way. For $C^+$, model (1), (3)–(5) is solved maximizing transportation costs and assuming all the potential warehouses are closed. Let $\Psi^+$ be the resulting maximum transportation cost, then $C^+ = \Psi^+ + \sum f_i$.

For the calculation of $C^-$, model (1), (3)–(5) is solved minimizing transportation costs and assuming that all the warehouses are open. Let $\Psi^-$ be the resulting minimum transportation cost, then $C^- = \Psi^- - \sum f_i$. For the calculation of $E^+$, model

**Fig. 2** Two membership functions used in the model

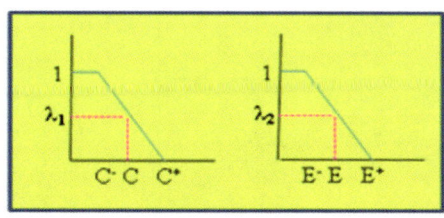

(2)–(5) is solved maximizing total emissions and assuming that all the warehouses are closed. Finally, for calculating $E^-$, model (2)–(5) is solved minimizing total emissions and assuming that all the warehouses open.

We assume that both objectives (minimizing total costs and total emissions) are equally important. As regards the constraints we shall request that their membership function values should be higher than a lower bound $\mu_{min}$ (see [11]). The model to solve is, thus, the following:

$$\max \lambda_1 + \lambda_2 \tag{6}$$

subject to

$$C = \sum_j f_j y_j + \sum_{i \in \hat{I}} \hat{c}_i \hat{t}_i \hat{x}_i + \sum_j \sum_{i \in I(j)} \left( \hat{\hat{c}}_j \hat{t}_j + c_{ji} t_{ji} \right) x_{ji} \tag{7}$$

$$E = \sum_{i \in \hat{I}} \hat{e}_i \hat{t}_i \hat{x}_i + \sum_j \sum_{i \in I(j)} \left( \hat{\hat{e}}_j \hat{t}_j + e_{ji} t_{ji} \right) x_{ji} \tag{8}$$

$$\lambda_1 \leq \frac{C^+ - C}{C^+ - C^-} \tag{9}$$

$$\lambda_2 \leq \frac{E^+ - E}{E^+ - E^-} \tag{10}$$

$$D_i^- + \mu \left( D_i^0 - D_i^- \right) \leq \hat{x}_i + \sum_{j:i \in I(j)} x_{ji} \leq D_i^+ + \mu \left( D_i^+ - D_i^0 \right) \quad \forall i \in \hat{I} \tag{11}$$

$$D_i^- + \mu \left( D_i^0 - D_i^- \right) \leq \sum_{j:i \in I(j)} x_{ji} \leq D_i^+ + \mu \left( D_i^+ - D_i^0 \right) \quad \forall i \notin \hat{I} \tag{11'}$$

$$\mu y_j \leq \frac{U_j^+ - \sum_{i \in I(j)} x_{ji}}{U_j^+ - U_j^-} y_j \quad \forall j \tag{12}$$

$$\mu y_j \leq \frac{\sum_{i \in I(j)} x_{ji} - L_j^-}{L_j^+ - L_j^-} y_j \quad \forall j \tag{12'}$$

$$0 \leq \lambda_1 \leq 1; \quad 0 \leq \lambda_2 \leq 1; \quad \mu_{min} \leq \mu \leq 1 \tag{13}$$

$$x_{ji}, \hat{x}_i \geq 0 \quad \forall i \in \hat{I}, \forall j \in I(j) \quad y_j \in \{0, 1\} \quad \forall j \tag{14}$$

## 4 Solution Procedure

In order to solve the above model a Genetic Algorithm (GA) will be used. The GA explores which warehouses are to be opened (binary variables $y_j$) and, for each individual, a Linear Programming (LP) solver is used to compute the corresponding fitness function selecting the best customer allocation, using model (6)–(14) with variables $y_j$ fixed (see Fig. 3). Note that, in principle, not every subset of warehouses is feasible, i.e., there is not always enough demand in the area of influence of the warehouses $I(j)$ to cover the minimum flow required to open the facilities as per constraints (12′). Therefore, a check needs to be done previous to calling the optimization software that solves the LP model. In case the candidate warehouses to be opened are seen to lead to an infeasible solution, changes in the warehouses subset are made until it can be assured there that the LP optimization software will return a feasible solution. This can be seen as a repair operator, which is one of the possible ways of handling constraints in GA.

Since the solution space explored by the GA corresponds to binary variables ($y_j$) a binary codification of the solution is used, i.e. each chromosome is just a vector of as many components as potential warehouse locations. Each component encode whether a warehouse is open or not. In order to assign a fitness value to an individual a linear solver is used to solve model (6)–(14) also obtaining the complete specification of the solution, including the flows between the central plant and the open warehouses and from these to their allocated customers.

About the crossover and mutation operators, standard binary coding operators have been used, namely the 1-point crossover (1X crossover) and the bitwise mutation. Fitness-proportional selection (i.e. roulette wheel) is used to choose the individuals to cross over. A generational GA is used with a maximum number of generations.

**Fig. 3** GA solution procedure solves LP model for fitness evaluation

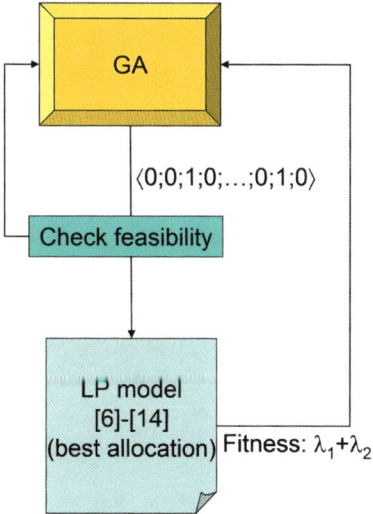

An additional stopping criterion consists in a limit on the number of generations without improving the best solution found.

As regards the implementation of the GA, an efficient parallel Python code has been programmed. Although the details of the parallelization strategy is out of the scope of this paper, let us just say that parallel python allows for calculating in parallel the fitness of all the individuals in initial population as well as of the new individuals created in each generation.

## 5 Computational Experiments and Results

For testing the good performance of the proposed approach, we have created a testbed of instances, each one with a 7 × 7 square grid of 48 potential warehouses locations (Fig. 4) and with the central plant in the middle of the grid (as used by [12]). The size of each of the grid cells is 125 × 125 km. The data were created in such a way that we have a clue about which could be the best possible solution, and then we shall check if our procedure is able to find a solution at least as good as that. With that purpose, customers were created locating them around a specific warehouse, forming a kind of cluster. Thus, for example, Fig. 5 shows an instance with four clusters of customers generated around four chosen warehouses. An additional cluster of customers, not in the vicinity of the four chosen warehouses, is also generated, with the expectation that these customers will likely be allocated to the central plant.

Two sets of 20 instances each were created. In the first set 2 warehouses are opened and 4 in the other. Therefore, 40 instances were solved and compared with the corresponding a priori "cluster" solution.

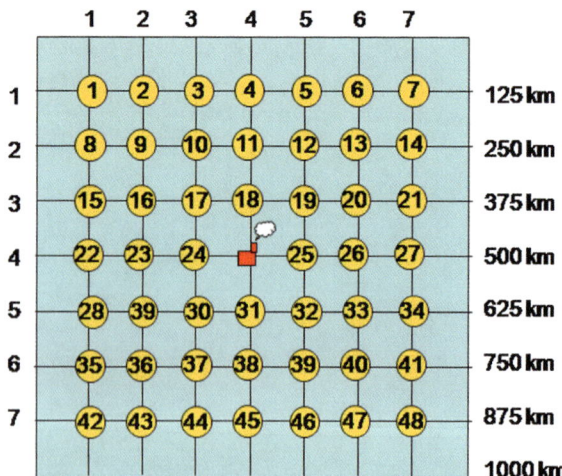

**Fig. 4** Basic layout of the generated instances

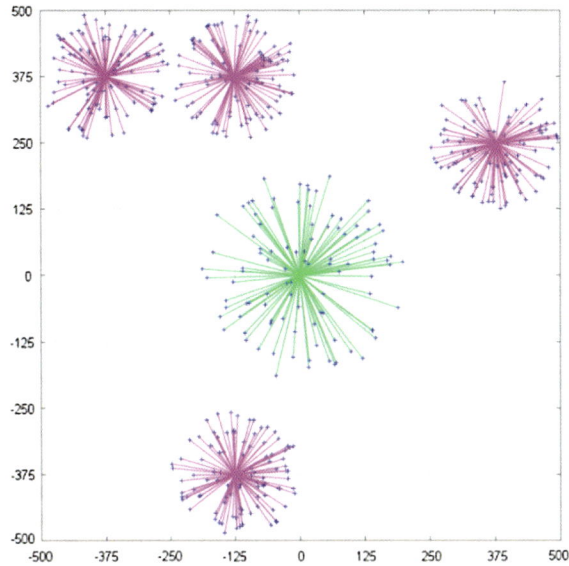

**Fig. 5** Example of a priori solution with 4 warehouses

**Table 2** Parameters for instance generation

| Type of vehicle assigned | Truck for central plant-warehouse movements and to serve a client with more than 15 ton/week of demand. Van: for the rest |
|---|---|
| Warehouse capacity | $K_j = 18750\,ton.\,U^- = 0.9 \cdot K_j;\, U^+ = 1.1 \cdot K_j$ |
| | $L^- = 0.52 \cdot U^+;\, L^+ = 0.6 \cdot U^-$ |
| Warehouse fixed cost | $F_j = 0.03 \cdot K_j$ |

For each of the $N = \{2, 4\}$ selected warehouses, a set of $(500/N)4/5$ clients in a radius distance of 125 km, all with the same demand, are randomly generated. The other fifth of the warehouse customers were generated out of that neighbourhood. Note that there is always a feasible solution since we assume that the central facility can always deliver goods to any client (although at a higher cost). Capacity and other parameters are assigned to each warehouse in such a way that the defined solution is feasible (Table 2). For the two types of vehicles (trucks and vans) cost and emission factors are shown in Table 3 and include the corresponding corrections to deal with non-full truckloads. The emission factors used correspond to those computed by the LIPASTO model developed by the Technical Research Centre of Finland (VTT) [13].

For the GA a population size of 100 was used, mutation probability was set to 0.001, maximum number of generations was 100 but stopping before reaching that limit if 10 generations pass without improving the best solution found.

**Table 3** Costs and emissions of trucks and vans depending on distance and load

|   | Truck | Van | Non-full truck, from central depot $> 125$ km | From warehouse $> 125$ km |
|---|---|---|---|---|
| $c_{ji}$ | 0.00004 €/kg/km | 0.00030 €/kg/km | | 0.00045 €/kg/km |
| $\hat{c}_i$ | 0.00004 €/kg/km | 0.00030 €/kg/km | 0.00006 €/kg/km | |
| $\hat{\hat{c}}_j$ | 0.00004 €/kg/km | 0.00030 €/kg/km | | |
| $e_{ji}$ | 0.0621 gr Eq-$CO_2$/kg/km | 0.0950 gr Eq-$CO_2$/kg/km | | 0.1425 gr Eq-$CO_2$/kg/km |
| $\hat{e}_i$ | 0.0621 gr Eq-$CO_2$/kg/km | 0.0950 gr Eq-$CO_2$/kg/km | 0.1425 gr Eq-$CO_2$/kg/km | |
| $\hat{\hat{e}}_j$ | 0.0621 gr Eq-$CO_2$/kg/km | 0.0950 gr Eq-$CO_2$/kg/km | | |

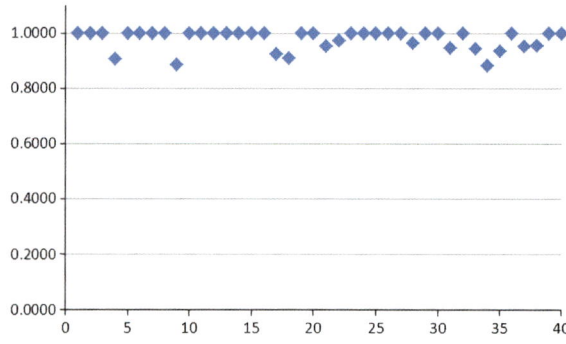

**Fig. 6** Ratio between the fitness of the final GA solution and the fitness of the original clustered solution

Comparing the results obtained with the clustered solution from which the instance customer data were generated, it can be seen in Fig. 6 that the GA procedure has been successful in 27 out of the 40 instances (two thirds of the cases) locating the warehouses according to the corresponding a priori clustered solution considered. Overall, the fitness of the GA solution (measured by $\lambda_1 + \lambda_2$) is 2.2 % below that of the a priori clustered solution. Note that as the problem complexity increases (as the number of clusters in the instance increases), it occurs more often (0 % in the case of two clusters, 45 % in 4 clusters case) that the GA does not find the a priori clustered solution (see Fig. 7). Different ways to compensate this effect are being studied to make the GA more robust.

For instance, in some trial experiments we have found that increasing the grain of the layout (considering 100 km instead of 125 km as distance between the marks in each axe, see Fig. 8), the results improve appreciably.

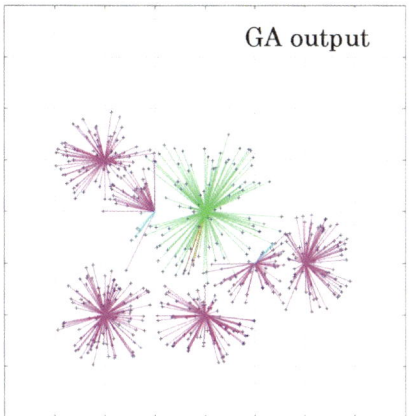
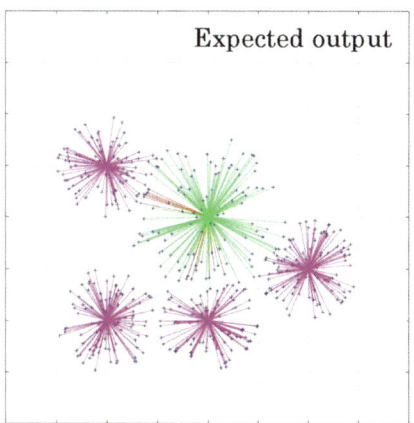

**Fig. 7** Example of cases when the GA output (*left*) was not able to match the expected solution (*right*)

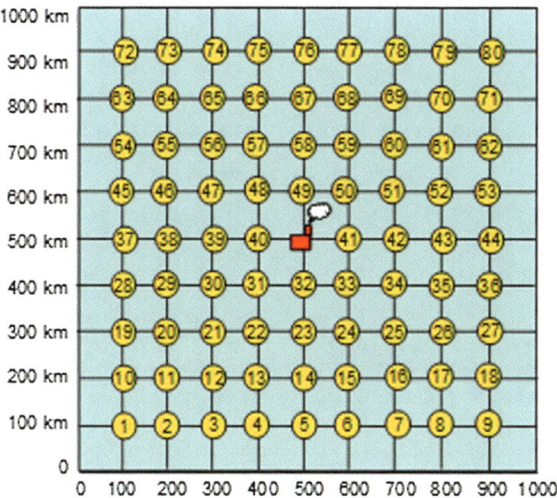

**Fig. 8** Reducing the grain the performance of the GA increases

# 6 Conclusions

This research has proposed a new network design approach that aims not only at cost minimization but also at minimizing GHG emissions from goods transportation. This second objective function will contribute to the sustainability of logistics operations. The decision variables are the selection of the warehouse location (from a set of discrete potential locations) and the allocation of customers to the selected warehouses.

A fuzzy bicriteria optimization model for solving the problem has been formulated and a GA solution procedure has been implemented. The GA explores the space of solutions corresponding to the selection of warehouses to open. A binary codification has been used so that if a potential location is opened the corresponding gene is one and zero otherwise. Standard crossover and mutation operator are used. Since there are both lower and upper bounds on the capacity of the open warehouses, a repair mechanism is needed to guarantee that these constraints hold and that the individual whose fitness is to be evaluated leads to a feasible solution.

A set of experiments have been carried out to check the performance of the procedure, using some instances for which a good reference solution is known a priori. The results indicate that the proposed approach often (although not always) finds this reference solution.

Once this tool for finding the most convenient assignment of clients to plant/warehouses is tested, as a further research the objective could be to check how including both objective functions affects the policy of designing centralized networks versus decentralized ones. A way to do that consists in preparing an experiment framework considering as factors the degree of concentration of clients, the number of clusters, and the number of clients. The dependent variable to analyze would be the percentage of clients directly served form the plant. Comparing the results using a single criterion versus using the bicriteria algorithm, would give clues about the influence of the environmental impact on centralization.

# References

1. Meixell, M.J., Gargeya, V.B.: Global supply chain design: a literature review and critique. Transp. Res. Part E. **41**, 531–550 (2005)
2. Pedersen, S.G., Zachariassen, F., Arlbjorn, J.S.: Centralisation versus de-centralisation of warehousing: a small and medium-sized enterprise perspective. J. Small Bus. Enterp. Dev. **19**(2), 352–369 (2012)
3. Abdul-Jalbara, B., GutiÂrrez, J., Puerto, J., Sicilia, J.: Policies for inventory/distribution systems: The effect of centralization versus decentralization. Intern. J. Prod. Econ. **81–82**, 281–293 (2003)
4. Chen, J.-M., Chen, T.-H.: The multi-item replenishment problem in a two-echelon supply chain: the effect of centralization versus decentralization. Comput. Oper. Res. **32**, 3191–3207 (2005)
5. Goh, M., Lim, J.Y.S., Meng, F.: A stochastic model for risk management in global supply chain networks. Eur. J. Oper. Res. **182**, 164–173 (2007)
6. Pishvaee, M.S., Razmi, J.: Environmental supply chain network design using multi-objective fuzzy mathematical programming. Appl. Math. Model. **36**, 3433–3446 (2012)
7. Chaabane, A., Ramudhin, A., Paquet, M.: Design of sustainable supply chains under the emission trading scheme. Intern. J. Prod. Econ. **135**(1), 37–49 (2012)
8. Akkerman, R., Farahani, P., Grunow, M.: Quality, safety and sustainability in food distribution: a review of quantitative operations management approaches and challenges. OR Spectrum. **32**, 863–904 (2010)
9. Melo, M.T., Nickel, S., Saldanha-da-Gama, F.: Facility location and supply chain management: a review. Eur. J. Oper. Res. **196**, 401–412 (2009)

10. Tiwari, R.N., Dharmar, S., Rao, J.R.: Fuzzy goal programming - an additive model. Fuzzy Sets Syst. **24**, 27–34 (1987)
11. Chen, L.H., Tsai, F.C.: Fuzzy goal programming with different importance and priorities. Eur. J. Oper. Res. **133**, 548–556 (2001)
12. Dasci, A., Verter, V.: A continuous model for production-distribution system design. Eur. J. Oper. Res. **129**, 287–298 (2001)
13. LIPASTO Unit emissions of vehicles, Freight transport - road traffic. http://lipasto.vtt.fi/ yksikkopaastot/tavaraliikennee/tieliikennee/tavara_tiee.htm

# Robustness Tools in Dynamic Dial-a-Ride Problems

**Samuel Deleplanque and Alain Quilliot**

**Abstract** The Dial-a-Ride Problem (DARP) models an operation research problem related to on demand transport. This paper introduces one of the fundamental features of this type of transport: the robustness. We solve the Dial-a-Ride Problem by integrating a measure of insertion capacity called *Insertability*. A greedy insertion algorithm integrating this measure is developed. We also use constraint propagation to manage all the time constraints (time windows, maximum ride times and maximum route times). The algorithm is able to measure the impact of each insertion on the other non-inserted demands. We study its behaviour, discuss the transition to dynamic context, and present a way to make the system more robust.

**Keywords** DARP · Robustness · Insertability · Dynamic · On-demand transportation

## 1 Introduction

Today, the Dial-a-Ride Problems are used in transportation services for elderly or disabled people. Also, the recent evolution in the transport field such as connected cars, autonomous transportation, and the emergence of shared services might need to use this type of problem at much larger scales. However, this type of transport is expensive and the management of the vehicles requires as much efficiency as possible. Moreover, the number of requests included in the vehicles planning can vary depending on the resolution used.

In [15], we solve the DARP by using constraint propagation in a greedy insertion heuristic. This technique obtains good results, especially in a reactive context, and is easily adaptable to a dynamic context. But, each demand is inserted one after another and the process does not take into account the impact of each insertion on the other

S. Deleplanque (✉) · A. Quilliot
LIMOS CNRS Laboratory, LABEX IMOBS3, 63000 Clermont-Ferrand, France
e-mail: deleplanque.samuel@gmail.com

A. Quilliot
e-mail: quilliot.alain@isima.fr

© Springer International Publishing Switzerland 2015
S. Fidanova (ed.), *Recent Advances in Computational Optimization*,
Studies in Computational Intelligence 580, DOI 10.1007/978-3-319-12631-9_3

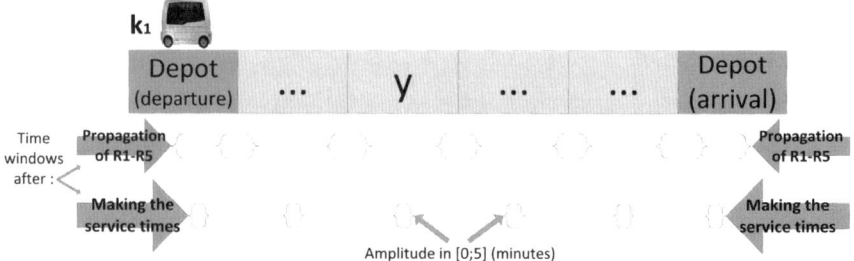

**Fig. 1** Times windows contraction

non-inserted demands, and so, in a dynamic context, the future demands. In this work, we present a measure of an insertion capacity named *Insertability*.

We introduce the *Insertability* calculation by indirectly integrating the impact of an insertion on the time constraints (time windows, maximum route time and maximum ride time) of the other demands.

This measure may be used in different steps of the resolution: selection of the demand to insert, selection of the insertion parameters, taking the decision to exclude a demand, and in the process making the service time. These four uses may be related to static as well as dynamic contexts. The main goal is to insert the current demand in a way which creates routes that will be able to insert future demands.

This paper is organized in the following manner: after a literature review, the next section will propose a model of the classic DARP. Then, we will review how to handle the temporal constraints with a heuristic solution based on insertion techniques using propagation constraints. We will continue by explaining the way to measure *Insertability*, a calculation based on the evolution of the time windows after an insertion. Then, we will give some other uses of this measure, including the setting of the service times which minimize the time windows (cf. Fig. 1). In the last part of the paper, the computational results will show the efficiency of our *Insertability*'s measure and we will report the evolution of the number of demands inserted in a resolution of some instances' sets.

## 2 Literature Review

The first work on the transportation optimization problem are related to the Traveling Salesman Problem [9]. Since that time, other transportation problems have emerged as the vehicle routing and scheduling problems, and also the Pick-up and Delivery Problem (PDP). Based on the PDP, the Dial-a-Ride problem has been studied since the 1970's. The DARP is related to the on demand transportation which is more often manage in a dynamic context, unlike the PDP.

There are a number of integer linear programmings [6], but the problem complexity is too high for using it in a real context. Indeed, it is an NP-Hard because it also generalizes the Traveling Salesman Problem with Time Windows (TSPTW). The optimal solution can be found only for small instances [8]. Therefore, and more specially for the dynamic context, the problem must be handled through heuristic techniques.

For the static DARP, [4] is an important study on the subject. In this work, they use the Tabu search metaheuristic to solve the problem. Starting from an initial solution, the resolution process moves from a solution to another in a neighborhood. The operator which executes these moves will take a demand already assigned to vehicle $k$ and insert it in another vehicle $k'$. The Tabu list saves each application of the operator in order to avoid cycling. The insertion parameters of the demand (for the origin and the destination) are chosen in order to minimize the total distance. They don't take into account all the constraints of the DARP during the process. They reach a feasible solution by integrating penalties in the performance criteria.

Other techniques are efficient for the static DARP. Dynamic programming (e.g. [3, 10]) or variable neighborhood searches (VNS) (e.g. [11, 12]) gave good solutions.

A basic feature of DARP is that it usually derives from a dynamic context. Therefore, algorithms for static DARP should be designed in order to take into account the fact that they will have to be adapted to dynamic and reactive contexts, which means synchronization mechanisms, interactions between the users and the vehicles, and uncertainty about forthcoming demands [7].

Reference [13], and later [14], developed the most used technique in dynamic context or in a real exploitation: heuristics based on insertion techniques. These techniques give good solutions even if the people's requests have to be taken into account in a very short period of time.

Reference [1] adapted the Tabu search of [4] to the dynamic DARP and a parallel environment. Each replication is independent in order to be executed on different CPU cores [1]. In fact, the operator moving from solution to another uses simple insertion technique. Other recent works have been done on the dynamic Dial-a-Ride problem, in particular about the problems of reactivity. For instance, in [2], they developed an algorithm that is able to take into account requests in real time. The requests are given directly to the driver, thus the decision is made immediately. Therefore, they describe the resolution process launched each time a new request appears in the system.

# 3 Mathematical Model

We first introduce the standard mathematical model for a single criterion DARP. The objective here is the minimization of the total distance. The DARP is defined by a transit network $G = (V, E)$, which contains at least two nodes *Depot* (departure and arrival), and whose arcs $e \in E$ are endowed with riding times equal to the Euclidean distance between two nodes of $V$, $i$ and $j$, $DIST(i, j) \geq 0$. A fleet $VH$ of $K$ vehicles with a capacity *CAP* and a Demand set $D = (D_i, i \in I)$. Any demand (or request) $D_i$ is defined as a 6-uple $D_i = (o_i, d_i, D_i, F(o_i), F(d_i), Q_i)$, where:

- $o_i \in V$ is the origin node of the demand $D_i$ and the set of all the $o_i$ is *DE*. $d_i \in V$ is the destination node of the demand $D_i$ and all the $d_i$ are gathered in *AR*;
- $F(o_i)$ and $F(d_i)$ are respectively the time windows for $o_i$ and $d_i$ of the demand $D_i$. *F.MIN* and *F.MAX* are the two bounds of a window;
- $\Delta_i \geq 0$ is an upper bound on the ride time of the demand $i$;
- $Q_i$ is the $D_i$'s load such that $q_{o_i} = Q_i = -q_{d_i}$.

We denote by $t_x^k$ the time at which the vehicle $k$ begins service in $x, x \in V$. Also, $\delta_j, j \in V$, is the non-negative service time necessary at the node $j$ and $\Delta^k$ is the maximum route time for the vehicle $k$.

Then, we consider in $G = (V, E)$ all the nodes corresponding to the $o_i \in V$ and $d_i \in V$ such that $V = DE \cup AR \cup \{0, 2|D| + 1\}$ with $\{0, 2|D| + 1\}$ the two depot nodes respectively for the departure and the arrival, $o_i \in \{1..|D|\}$, and $d_i \in \{(|D| + 1)..(2|D|)\}$. Moreover we denote by $V'$ the set $V$ without the depot nodes (i.e. $V' = DE \cup AR$) and by $\zeta_j^k$ the total load of the vehicle $k$ leaving the node $j, j \in V$.

Dealing with such an instance means planning the handling demands of $D$, by the fleet *VH*, while taking into account the constraints which derive from the technical characteristics of the network G, of the vehicle fleet *VH*, and of the 6-uples $D_i = (o_i, d_i, D_i, F(o_i), F(d_i), Q_i)$, and while optimizing some performance criterion may is usually minimizing the total distance or a mix of an economical cost (point of view of the fleet manager) and of QoS criteria (point of view of the users).

Let $x_{ij}^k$ a boolean equals to 1 if the vehicle $k$ travels from the node $i$ to the node $j$. Then, based on [5], the mathematical formulation is the following mixed-integer program :

$$Min \sum_{k \in K} \sum_{i \in V} \sum_{j \in V} DIST(i, j) x_{ij}^k \tag{1}$$

subject to

$$\sum_{k \in K} \sum_{j \in V'} x_{ij}^k = 1, \quad \forall i \in V' \tag{2}$$

$$\sum_{j \in V'} x_{ij}^k - \sum_{j \in V'} x_{|D|+i, j}^k = 0, \quad \forall i \in DE, \quad k \in K \tag{3}$$

$$\sum_{j \in DE} x_{0j}^k = 1, \quad \forall k \in K \tag{4}$$

$$\sum_{j \in V'} x_{ji}^k - \sum_{j \in V'} x_{ij}^k = 0, \quad \forall i \in V', k \in K \tag{5}$$

$$\sum_{i \in AR} x^k_{i,2|D|+1} = 1, \quad \forall k \in K \tag{6}$$

$$t^k_j \geq (t^k_i + \delta_i + DIST(i, j))x^k_{ij}, \quad \forall i \in V, j \in V, \quad k \in K \tag{7}$$

$$\zeta^k_j \geq (\zeta^k_i + q_j)x_{ij}, \quad \forall i \in V, \quad j \in V, \quad k \in K \tag{8}$$

$$DIST(i, |(D)| + i) \leq t^k_{|(D)|+i} - (t^k_i + \delta_i) \leq \Delta_i, \quad i \in DE \tag{9}$$

$$F.MIN_i \leq t^k_i \leq F.MAX_i, \quad \forall i \in V, k \in K \tag{10}$$

$$t^k_{2|D|+1} - t^k_0 \leq \Delta^k, \quad \forall k \in K \tag{11}$$

$$\zeta^k_i \leq CAP^k \tag{12}$$

$$x^k_{ij} \in \{0; 1\}, t \in R^+ \tag{13}$$

The program above is a three index formulation (report to [5] for more explanations about the objective function (1) and the constraints (2)–(13)). Several other mathematical formulations for the DARP exist, even some with a two index formulation [6]. But, the complexity of all these linear programs does not allow for finding an exact solution with a solver, the operation is too time consuming.

Throughout this work, we will to deal with homogeneous fleets and nominal demands, and we shall limit ourselves to static points of view but our insertion process allows flexibility for using it in a dynamic context. Still, we shall pay special attention to cases when temporal constraints are tight.

## 4 A Greedy Insertion Algorithm

This section explains the techniques used to solve a static DARP instance. In our framework, the routes are saved in lists $\Gamma_k$ surrounded by two depot nodes (one for the departure on the other for the arrival) with $k$ a vehicle of the fleet $VH$ (cf. Fig. 1). For any sequence (or list) $\Gamma_k$ we set:

- for any $z$ in $\Gamma_k$, $\text{Succ}(\Gamma_k, z) = $ Successor of $z$ in $\Gamma_k$ and $\text{Pred}(\Gamma_k, z) = $ Predecessor of $z$ in $\Gamma_k$ ;
- for any $z, z'$ in $\Gamma_k$, $z \ll_k z'$ if $z$ is located before $z'$ in $\Gamma_k$ and $z \ll^=_k z'$ if $z \ll_k z'$ or $z = z'$.

We also defined the $Twin$ function such that, for any node $o_i$ (respectively $d_i$) which appears in $\Gamma_k$, the node $Twin(o_i) = d_i$ (and $Twin(d_i) = o_i$). The twin of a depot node is the other depot node in the same list.

In [15], we present an insertion greedy algorithm based on constraint propagation in order to contract time windows according to the time constraints of the DARP.

An insertion which does not imply constraint violation is said *valid* if $\Gamma = \cup_{k \in K} \Gamma_k$, the resultant collection of routes, if *load-valid* and *time-valid*. A route is *load-valid* if the capacity is not exceed, so, the *load-validity* is obtained if $ChT_k(x) \leq CAP$ with $ChT_k(x) = \sum_{y \ll_k^= x} q_y$, $x$ and $y$ nodes in the route $k$. The *time-validity* is obtained if there is no violation of the time constraints modeling by, for each demand $i$, $i \in D$, $\Delta_i$ the maximum ride time, $\Delta^k$, $k \in K$ the maximum route time and the constraints modeled by each time window $F(o_i) = [F.min(o_i), F.max(o_i)]$ and $F(d_i) = [F.min(d_i), F.max(d_i)]$. The service time $\delta_j$ of each node $j \in V$ and origin of an arc $e \in E$ is added in the DIST matrix, but without considering the departure depot nodes. Checking the *load-validity* on $\Gamma = \cup_{k \in K} \Gamma_k$ is easy, and we show the efficiency of the constraint propagation in order to prove to *time-validity* after each planned insertion once the *load-validity* is proved. According to a current time window set $FP = \{FP(x) = [FP.min(x), FP.max(x)], x \in \Gamma_k, k = 1..K\}$ the *time-validity* may be performed through propagation of the five following inference rules **Ri**, i = 1..5 in a given route $\Gamma_k$:

for each $(x,y)$ pair of nodes such that $y = Succ(\Gamma_k, x)$:

- **R1**:
$$FP.min(x) + DIST(x, y) > FP.min(y)$$
$$\models$$
$$(FP.min(y) \leftarrow FP.min(x) + DIST(x, y)); (FNodeModif \leftarrow y)$$

- **R2**:
$$FP.max(y) - DIST(x, y) < FP.max(x)$$
$$\models$$
$$(FP.max(x) \leftarrow FP.max(y) - DIST(x, y)); (FNodeModif \leftarrow x)$$

if $(y = Twin(x))$ and $(x \ll_\Gamma y)$:

- **R3**:
$$FP.min(x) < FP.min(y) - \Delta_i$$
$$\models$$
$$(FP.min(x) \leftarrow FP.min(y) - \Delta_i); (FNodeModif \leftarrow x)$$

- **R4**:
$$FP.max(y) > FP.max(x) + \Delta_i$$
$$\models$$
$$(FP.max(y) \leftarrow FP.max(x) + \Delta_i); (FNodeModif \leftarrow y)$$

and for each $x$, $x \in \Gamma_k$, $k = 1..K$:

- **R5**: $FP.min(x) > FP.max(x) \models REJET \leftarrow true$.

These 5 rules are propagated in a loop while there no time windows exists $FP$ modified at the last iteration, i.e. once $FNodeModif$ is empty. In the R3 and R4 rules, the '$\Delta_i$' is replaced by '$\Delta_k$' if the pair of nodes $(x,y)$ are depots. The tour $\Gamma_k$, $k = 1..K$, is *time-valid* according to the input time window set $FP$ if and only if the *REJET* Boolean value is equal to *false* as initialized at the beginning of the process. In such a case, any *valid-time* value set $t$ related to $\Gamma_k$ and $FP$ is such that: for any $x$ in $\Gamma_k$, $t(x)$ is the service time in $FP(x)$.

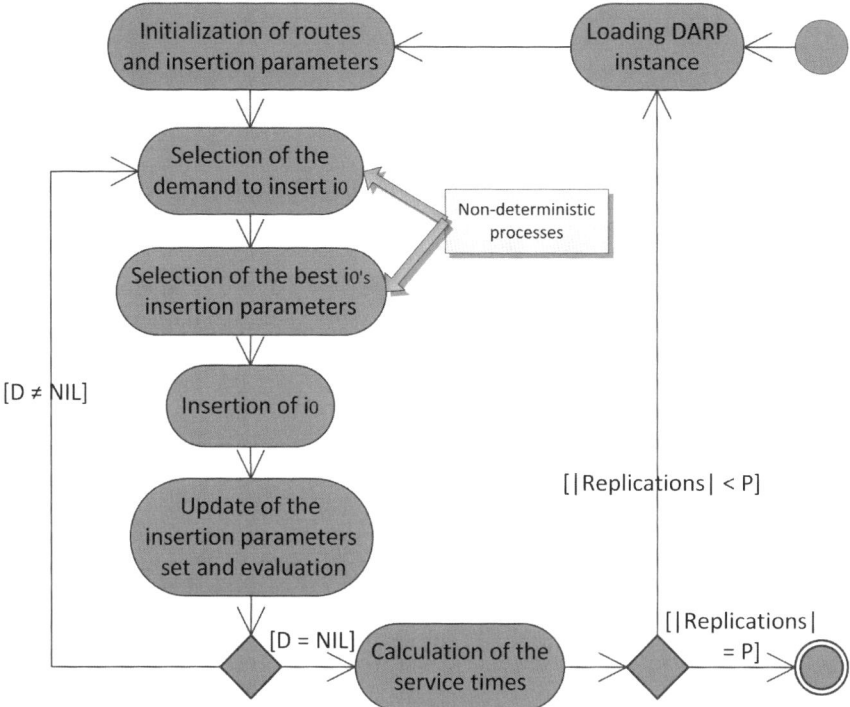

**Fig. 2** The Monte-Carlo process

The greedy insertion algorithm includes this propagation constraint technique in order to evaluate each possible insertion. Each iteration of the algorithm selects one demand according to the number of vehicle able to integrate it. Once a demand is selected, the process chooses the best insertion's parameters that are the vehicle and the location of the *origin* and *destination* nodes. These parameters are chosen according to the smaller evolution of the routes costs (like the total distance in the mathematical model). The selection of the demand and the parameters are made in a non-deterministic way (among a small set of the best variables). Then, we are allowed to include our process in a Monte Carlo scheme. The Fig. 2 summarizes the general algorithm with $P$ the number of replications resolving a DARP instance.

## 5 Insertability Optimization

In the above algorithm, each iteration selects a demand, and then, it finds the way to insert while minimizing the total cost. This greedy algorithm does not take into account the total impact of this actual insertion on the future demands integration, but only the effect on the demands already inserted. In this section, we introduce a

*Insertability* calculation by integrating this impact of an insertion related to the time constraints (time windows, maximum ride time and maximum route time).

During the insertion process, the state of the system is given by:

- a set of demands $D - D1$ already integrated in the routes, and $D1$ is the set of demands not inserted,
- a collection $\Gamma = \cup_{k \in K} \Gamma_k$ of routes including a list of nodes related to the *Depot*, *origin* and *destination* nodes,
- a exhaustive list of insertion's parameters sets. Each set gathers 5 elements : $k$ the vehicle, $i$ the demand, $(x, y)$ the pair of insertion nodes (locating respectively $o_i$ between $x$ and the successor of $x$, and $d_i$ between $y$ and the successor of $y$), and $v$ the evolution of the collection $\Gamma = \cup_{k \in K} \Gamma_k$ 's cost.

## 5.1 The Insertability Measure

Given that the difficulty of the instances is linked to the time constraints, we introduce an *Insertability* calculation related to the times windows contractions. During an insertion's assessment, these reductions appear once the inference rules are propagated. Here, we try to find a good triple $(k, x, y)$, the vehicle and the location of the *origin/destination* nodes, in order to give enough "time space" to the future demands (which have to be integrated in $\Gamma = \cup_{k \in K} \Gamma_k$).

We set  *INSER(i, $\Gamma$)* the *Insertability* measure of the demand $I$. The quantity $U_n^k(z)$ denotes the vehicle $k$ time windows' amplitude of the node $n$ once it has been inserted to the right of node $z$. *INSER* is calculated as follows:

- $INSER(i, \Gamma) = \sum_{k \in K} INSER1(i, \Gamma_k)$ ;
- $INSER1(i, \gamma) = Max_{(x,y)} INSER2(i, \gamma, x, y)$, $\gamma$ a tour of $\Gamma$ ;
- $INSER2(i, \gamma, x, y) = U_{o_i}^{\gamma}(x).U_{d_i}^{\gamma}(y)$.

*INSER2* is the product of the time windows amplitudes resulting to the insertion of the demand $i$ in the tour $\gamma$. The origin $o_i$ is inserted to the right of $x$ and the destination $d_i$ to the right of $y$. The product was chosen in order to penalize the *Insertability* if the time windows of the origin and/or the destination are tight and also to have *INSER2* equals to zero if one of the two nodes can not be inserted.

*INSER1* gives us the maximum of *INSER2* over the possible insertion positions $x$ and $y$ in the route $\gamma$. When *INSER1* is equal to 0, the new route $\gamma$ resulting to the new insertion is not *time-valid*.

Finally, *INSER* makes the sum of all the *INSER2* for each vehicle of the fleet $VH$. If *INSER* equals to zero, the demand can not be inserted in all cases.

We set *Inserted($\Gamma$, $i_0$, $k$, $x$, $y$)* the updated collection of tours $\Gamma$ with the insertion of the selected demand $i_0$ at the locations $x$ and $y$ in the vehicle $k$.

The *INSER(i, $\Gamma$)* measure allows us to write the *Optimization Insertability Problem* which consists to find the best insertion parameters in order to keep the vehicles' scheduling more flexible (in a sense to let enough possibilities of insertions for the future demands):

## 5.2 Optimization Insertability Problem

Find the optimal parameters $(k,x,y)$ inserting $i_0$ and maximizing the value $Min_{i \in D1-i_0} INSER(i, Inserted(\Gamma, i_0, k, x, y))$.

The value $Min_{i \in D1-i_0} INSER(i, Inserted(\Gamma, i_0, k, x, y))$ may be used if all the demands have to be inserted. Another optimization may be process as the maximization of the sum $\sum_{i \in D1-i_0} INSER(i, Inserted(\Gamma, i_0, k, x, y))$. The choice is made according to the homogeneity of the demands and if the problem requires to insert all the set $D$.

This problem only optimizes the variation of the *Insertability* values and does not include other performance (or costs) criteria like the minimization of the ride times, waiting times or distances. The *Insertability* criterion can be integrated in a mix of economical cost (point of view of the fleet manager) and of QoS criteria (point of view of the users). Then, the process maximizes the function $Perf = \mu \cdot \sum_{i \in D1-i_0} INSER(i, Inserted(\Gamma, i_0, k, x, y)) - v(Inserted(\Gamma, i_0, k, x, y))$ with $\mu$ a criterion coefficient and $v$ the cost value function mixing the costs related to the both points of view.

## 5.3 Other Uses of the Insertability Measure

So far, we select the demand $i_0$ according to the number of vehicles available (taking in account all the time and load constraints). The *Insertability* measure $INSER(i_0, \Gamma)$ may be also used in order to select the next request $i_1$ to insert. This application could be used in a context where all the demands of $D$ have to be integrated. The selection is based on the smallest *Insertability* measure. Once a demand is selected, the problem may solve the *Optimization Insertability Problem*. Here, the two steps may be written in a non-deterministic way. The demand may be selected randomly through a set of N1 elements with the smallest *INSER* value. The same scheme may be applied on a set of a insertion parameters of N2 elements with a best $(k, x, y)$ elements maximizing the quantity $Min_{i \in D1-i_0} INSER(i, Inserted(\Gamma, i_0, k, x, y))$.

Also, $INSER(i_0, \Gamma)$ may be useful for a larger set $D$. If the instance does not have any solution integrating all the set $D$, it is preferable to identify requests to exclude as soon as possible. The exclusion of a demand $i_0$ may be set up if its insertion results in $\Gamma$ not enough flexible to include the other elements of $D1$. In other words, the demands excluded will be those that will have the most impact of future insertions. The difference $\sum_{i \in D1-i_0} (INSER(i, \Gamma) - INSER(i, Inserted(\Gamma, i_0, k, x, y)))$ of the inequality (14) takes in account the *Inserability* measure of $D1 - i_0$ before and after the insertion of $i_0$ in the routes of $\Gamma$. If this difference is larger than the threshold $\xi$, the demand is excluded. In the experimentation section, we will discuss the fact this threshold should be dynamic and decreases over the execution.

$$\sum_{i \in D1-i_0} (INSER(i, \Gamma) - INSER(i, Inserted(\Gamma, i_0, k, x, y))) > \xi \qquad (14)$$

## 5.4 The Insertability Optimization Suited to the Greedy Insertion Algorithm

The calculation of $INSER(i, \Gamma)$, $i \in D$, begins to be time consuming starting from a medium size of $D$, while the $INSER2$ value is based on the time windows amplitude obtained after the propagation of the time constraints. Therefore, this is important to spot each step of the process where the *Insertability* measure does not have to be updated. When $i_0$ is selected, $INSER2(i, \Gamma_k, x, y)$, $INSER1(i, \Gamma_k)$ and $INSER(i, \Gamma)$ are known for all demand in $D1 - i_0$ and all $k = 1..K$. Once $i_0$ is about to be inserted, the process computed the value $H(i)$, $i \in D1 - i_0$ (cf. formulation (15)). Then, the algorithm tries the insertion of each $i$ from $D1 - i_0$ in $Inserted(\Gamma, i_0, k, x, y)$ and deduces the value $K(i)$ given in formula (16) for all $i \in D1 - i_0$ and, ultimately, the quantity $Val(k, x, y) = Min_{i \in D1 - i_0}(K(i) + H(i))$.

$$H(i) = INSER(i, \Gamma) - INSER1(i, \Gamma_k) \tag{15}$$

$$K(i) = INSER(i, Inserted(\Gamma, i_0, k, x, y))$$
$$= H(i) + INSER1(i, Inserted(\Gamma, i_0, k, x, y)_k) \tag{16}$$

Other calculations may be avoided. First, we set the variable $W_1$ such that $W_1 = Min_{i \in D1 - i_0} INSER(i, \Gamma)$. If the quantity $INSER(i, \Gamma) - INSER1(i, \Gamma_k)$ is larger than $W_1$, there is no need to test the impact of the insertion of $i_0$ on $i$.

Finally, we are able to use $INSER(i, \Gamma)$ once we integrate the future demands presented in the next section. In a dynamic context, the *Insertability* measure helps the routes to be enough flexible for the next insertion process. Moreover, the service time have to be set with the same purpose and $INSER(i, \Gamma)$ is able to help to do it.

## 6 Future Demands in the Dynamic DARP

The problem may have to be handled in a dynamic way and the greedy insertion algorithm is easily adaptable to this context. Once the *Insertability* measure is included in the performance criteria, the system may increase its robustness and we need to exploit knowledge about future demands for that. In our case, this knowledge is related to the type of on demand transportation service. In this paper, we will use a simple extrapolation of this probable demand based on the demand already broadcasted.

We will not take into account the way the system supervises its various communication components with the users. In reality, there are eventual divergences between the data which were used during the planning phases and the situation of the system.

We set $D$-$V$ the virtual demands, $D$-$R$ the real demands, and $D$-*Rejet* the set of the ones excluded from the insertion algorithm. The $D$-$V$ formulation is given in (17). $p_i$ gives us the number of times the future demand $D_f$ will appear for each period of each discrete planning horizon.

$$D\text{-}V = \sum_{i \in D_f} D_i.p_i \qquad (17)$$

Then, we are able to update the formula (18) giving the performance function *Perf*.

$$Perf = \alpha. \sum_{i \in D_f} p_i INSER(i, Inserted(\Gamma, i_0, k, x, y))$$

$$+ \mu. \sum_{i \in D1 - i_0} INSER(i, Inserted(\Gamma, i_0, k, x, y)) - v(Inserted(\Gamma, i_0, k, x, y))$$

$$(18)$$

As in the previous sections, the process may exclude some demands taking in account the future requests. We updated the inequality (14) by the (19). $\alpha$ is a coefficient based on the importance of the future demands.

$$\alpha. \sum_{i \in D_f} p_i.(INSER(i, \Gamma) - INSER(i, Inserted(\Gamma, i_0, k, x, y)))$$
$$+ \sum_{i \in D1 - i_0} (INSER(i, \Gamma) - INSER(i, Inserted(\Gamma, i_0, k, x, y))) > \xi \quad (19)$$

# 7 Discussion About the Service Times and the Dynamic Context

Most works on vehicle scheduling problems including time window studies how to integrate a set of demands in the vehicle planning. Making a service time anticipating the future is especially rare.

Once routes are built and integrated a first set $D$, the users expect the date when the vehicle selected will pick them up. In the lists forming the $K$ routes, each node has a time window. After the service time is set, each time window becomes tight with zero amplitude or equals a very small delay. How the service times are made is very important for the next insertion's process.

For instance, we consider a fleet of 2 vehicles with two schedules made by a first resolution of the DARP minimizing the total distance. The date of each service time has already given to the users. Once these dates are created, the system has to reduce the time windows to 0 like in the Fig. 3. While the vehicles are not already left from the depots, the system needs to integrate one more demand in a second resolution process. Then, each new demand such as (o6,d6) will not be able to be inserted.

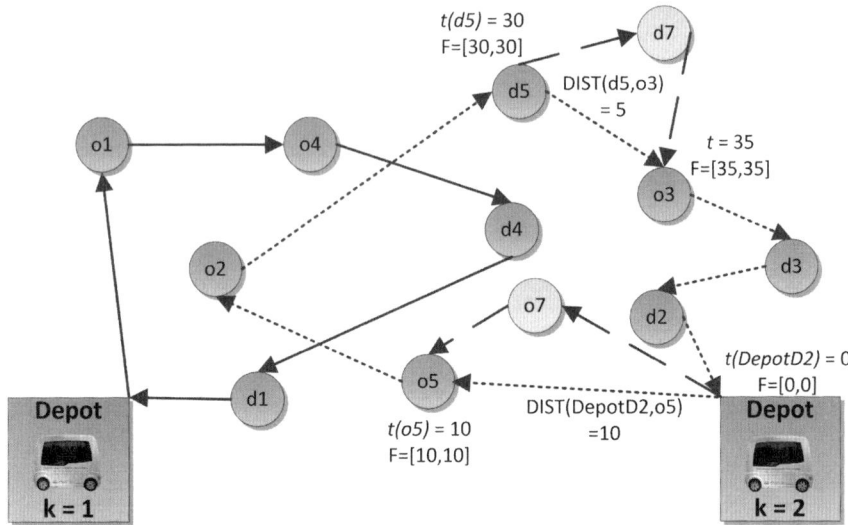

**Fig. 3** New insertions after the first set of service times

Indeed, for the origin node we have: $t(Depot\,D2) + DIST(Depot\,D2, o5) = t(o5)$ which make the insertion of o6 impossible.

One more time, the $INSER(i, \Gamma)$ values may be used in order to set the service time without to have the problem above. The service time may be calculated once the process have inserted the virtual demands $D\text{-}V$ and the real demands $D\text{-}R$.

The previous section shows the way to anticipate the future demands $D\text{-}V$. These demands are related to a dynamic context. Note again that our greedy algorithm is easily adaptable to this context. More specifically, the technique does not change unlike the state of each route. The first node is not a depot node anymore but a dynamic node related to the vehicle's location. The entire constraint propagation process is applied on these new routes. A simulation will be necessary to evaluate the anticipation of the future demands including in the dynamic context.

## 8 Computational Experiments

In this section, we study the behavior of our *Insertability* measure used in the resolution of Dial-a-Ride instances. The algorithms were implemented in C++ and compiled with GCC 4.2. In [15], we solve the [4]'s instances by our greedy insertion algorithm based on constraint propagation. We obtained good results in the majority of instances, but, only 1 % of the replications gave us a feasible solution on the tenth instance (*R10a*). The CPU time was smallest or equal to the best times in the literature; we do not work on this feature for this experiment.

## 8.1 First Experimentation: The Optimization of the Selection of the Demand to Insert

### 8.1.1 INSER Used in the Selection of a Demand

We note by $R^{DARP}$ the rate of 100 replications which give us a feasible solution obtained by using the solution of [15]. Here, the selection of the demand is based on the lowest number of cars which are able to accept it. $R_{Rob}^{DARP}$ is the rate obtained with the same process except that each demand is selected at each iteration by the lowest *Insertability* value *INSER*.

The *Insertability* measure is already efficient once it's used in the selection of the demands to insert. The rate obtained for the pr08, pr09, pr10 and pr19 are clearly more interesting as shown in Table 1 (e.g. for the instance pr08, the rate increases by 56 to 91 % of success).

**Table 1** $R^{DARP}$ vs $R^{DARP Rob}$

| Inst. | $R^{DARP}$ | $R_{Rob}^{DARP}$ |
|-------|-----------|-----------------|
| pr01  | 99        | 100             |
| pr02  | 100       | 100             |
| pr03  | 97        | 100             |
| pr04  | 100       | 100             |
| pr05  | 100       | 100             |
| pr06  | 100       | 100             |
| pr07  | 90        | 96              |
| pr08  | 56        | **91**          |
| pr09  | 18        | 21              |
| pr10  | 1         | **7**           |
| pr11  | 100       | 100             |
| pr12  | 100       | 100             |
| pr13  | 99        | 100             |
| pr14  | 100       | 100             |
| pr15  | 100       | 100             |
| pr16  | 100       | 100             |
| pr17  | 98        | 100             |
| pr18  | 99        | 100             |
| pr19  | 64        | **99**          |
| pr20  | 43        | 56              |
| Av.   | 83.2      | 88.5            |

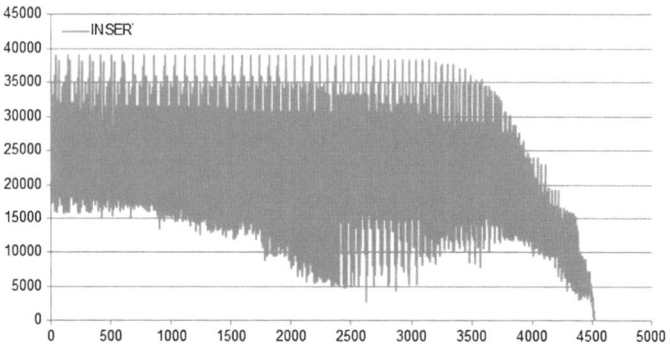

**Fig. 4** INSER values on the not inserted demands

### 8.1.2 INSER Behaviour

Each time a replication cannot integrate all the request, the *INSER* value of the demands not inserted has to be null. In Fig. 4, while the resolution process applied to the R10a instance, we note the evolution of more than 4,500 *INSER*'s demands not inserted. The technique used is the second approach selecting the demand by the smallest *Insertability*. The values noted are from a failed replication.

One can observe big gaps between the different *INSER* until the 4,000 first values. After that, for the remaining requests, the *Insertability* values decrease strongly because the routes begin to be not flexible. Between the $2,500$th and the $3,500$th, for some demands, the values are very low at the beginning just before increasing strongly. This is explained by the fact the process inserts the demand with the lowest *INSER* but their insertion do not make a big impact on the other demands not inserted. This impact is related to the *Optimization Insertability Problem* studied below.

## 8.2 Second Experimentation: The Optimization of the Insertion Parameters

In a second experimentation, we compare the [15]'s approach and another algorithm based on the optimization of the parameters $(x,y,k)$. The selection of the request to insert is the same for both solutions. For the second one, once a demand $i_0$ is selected, we maximize the sum $\sum_{i \in D1-i_0} INSER(i, Inserted(\Gamma, i_0, k, x, y))$ in order the find the best parameter $(x,y,k)$ which will integrate $i_0$ in the route $k$. We do not optimize $Min_{i \in D1-i_0}INSER(i, Inserted(\Gamma, i_0, k, x, y))$ because we create instances especially with a set $D$ too large for inserting all the requests. Therefore, the demand with the smallest value *INSER* for a given parameters $(x,y,k)$ could never be integrated into the routes.

**Table 2** Parameters' instances

| K | $e_{F(o)}$ | $e_{F(d)}$ | $\Delta$ | $CAP$ |
|---|---|---|---|---|
| 10 | 35 | 10 | $\infty$ | 10 |

**Table 3** Gap between the INSERT rates

| $|D|$ | 50 | 75 | 100 | 150 | 200 |
|---|---|---|---|---|---|
| $T_{Insert}$ | 100 | 93.2 | 78.9 | 64.2 | 52.6 |
| $T_{Insert_{Rob}}$ | 100 | 96.8 | 85.3 | 66.4 | 54.1 |
| $Gap_{Insert}$ | 0 | 3.86 | 8.11 | 3.43 | 2.81 |

The two algorithms were applied to five sets of 5 randomly generated instances. All the instances have their time constraints related to the interval [0;400] and all the load was unit. We set by $e_{F(o)}$ and $e_{F(d)}$ the amplitude of the time windows at the *origin* and the *destination* given by the users, respectively. The other parameters are given in Table 2.

We generate 5 different sets of 5 instances with a variation of the number of demands $|D|$. We set by $T_{Insert}$ and by $T_{Insert_{Rob}}$ the demand inserted's rate the first resolution and the second technique, respectively. Finally, $Gap_{Insert}$ is the gap in percentage between each rate. Its calculation is given by $Gap_{Insert} = 100 \cdot (T_{Insert_{Rob}} - T_{Insert})/T_{Insert}$. We launched 100 replications of each technique on the 5 sets. The results are provided by the Table 3.

In future work, we could integrate the usual performance criterion (like the total distances) in the *Perf* value in order to choose each best insertion parameters. Here, we are just taken into account the *INSER* values in order to integrate the most requests possible. The results show us that the larger of $|D|$ defines if the system needs to optimize the *Insertability* measure. For $|D| = 50$, all the requests are able to be inserted easily, so, the *INSER* values does not have any interest. When the set is composed of 100 demands, we obtained a $Gap_{Insert}$ of 8.11 % meaning there are more than 8 % more requests inserted by the second approach.

For this set of instance, we also tried to integrate a new feature in our algorithm: we've added the ability to exclude a request if the impact of one insertion involving a significant drop of the general *Insertability*'s demands from $D1 - i_0$. Before that, we study the threshold which limits the variation of *Insertability*.

$i_0$ is excluded if $\sum_{i \in D1-i_0}(INSER(i, \Gamma) - INSER(i, Inserted\ (\Gamma, i_0, k, x, y)))$ $> \xi$ is true with $\xi$ a threshold. The calculation of the threshold is not easy. In the Fig. 5, we report *Variation* which is the difference $INSERav - INSERap$ with *INSERav* and *INSERap* the values $\sum_{i \in D1-i_0} INSER(i, \Gamma)$ and $\sum_{i \in D1-i_0} INSER(i, Inserted\ (\Gamma, i_0, k, x, y))$, respectively. This figure shows us that the threshold $\xi$ have to be calculated dynamically according to the average of *INSER*.

We used this type of dynamic threshold for the third set of instances with 100 demands. We exclude an request if the current $\xi$ is exceeded, and only this feature is added in the second approach. We obtained a gain of 1,3 % in average (from 85,3 % to 86,6 %) meaning approximately one more demand is able to be inserted.

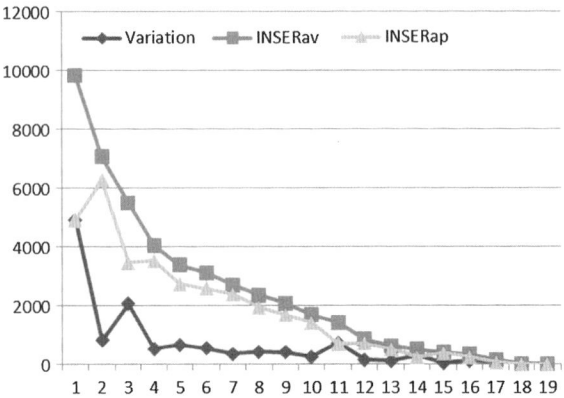

**Fig. 5** Variation of the *Insertability* values between each insertion

## 9 Conclusion

The Dial-a-Ride Problem is one of the transport problems with the highest number of hard constraints. The insertion techniques are able to obtain a good solution in a reasonable amount of time. Their adaptability to a dynamic context is easy, could lack robustness if the goal is to integrate as many requests as possible.

We have introduced a way to measure the impact of each insertion on the other demands which have not yet been inserted. This *Insertability* measure could be used in order to exclude a demand, to select a demand to insert and also to calculate the best insertion parameters. This value, named *INSER*, leads to a large amount of work opportunities. We have introduced a simple way to model the future demands, and how to adapt our greedy insertion algorithm based on the constraint propagation to the dynamic context. In future work, we will develop a simulation which is necessary to evaluate the efficiency of our techniques in a simulation. The final goal will be to adapt these techniques to a real case.

**Acknowledgments** This work was founded by the French National Research Agency, the European Commission (Feder funds) and the Region Auvergne in the Framework of the LabEx IMobS3.

## References

1. Attanasio, A., Cordeau, J.-F., Ghiani, G., Laporte, G.: Parallel Tabu search heuristics for the dynamic multi-vehicle dial-a-ride problem. Parallel Comput. **30**(3), 377–387 (2004)
2. Coslovich, L., Pesenti, R., Ukovich, W.: A two-phase insertion technique of unexpected customers for a dynamic dial-a-ride problem. Eur. J. Oper. Res. **175**(3), 1605–1615 (2006)
3. Chevrier, R., Canalda, P., Chatonnay, P., Josselin, D.: Comparison of three algorithms for solving the convergent demand responsive transportation problem, In: 9th International IEEE Conference on Intelligent Transportation Systems, pp. 1096–1101. Toronto, Canada (2006)

4. Cordeau, J.-F., Laporte, G.: A tabu search heuristic algorithm for the static multi-vehicle dial-a-ride problem. Transp. Res. B **37**, 579–594 (2003)
5. Cordeau, J.-F.: A branch-and-cut algorithm for the Dial-a-Ride. Oper. Res. **54**, 573–586 (2006)
6. Cordeau, J.F., Laporte, G.: The dial-a-ride problem: models and algorithms. Ann. Oper. Res. **153**(1), 29–46 (2007)
7. Ehmke, J.F.: Integration of Information and Optimization Models for Routing in City Logistics. International Series in Operations Research and Management Science. Springer, New York (2012)
8. Gendreau, M., Potvin, J.Y.: Dynamic vehicle routing and dispatching. In: Crainic, T.G., Laporte, G. (eds.) Fleet Management and Logistics, pp. 115–126. Kluwer, Boston (1998)
9. Menger, K.: Das botenproblem. Ergebnisse eines mathematischen kolloquiums **2**, 11–12 (1932)
10. Psaraftis, H.: An exact algorithm for the single vehicle many-to-many dial-a-ride problem with time windows. Transp. Sci. **17**, 351–357 (1983)
11. Parragh, S.N., Doerner, K.F., Hartl, R.F.: Variable neighborhood search for the dial-a-ride problem. Comput. Oper. Res. **37**, 1129–1138 (2010)
12. Healy, P., Moll, R.: A new extension of local search applied to the dial-a-ride problem. Eur. J. Oper. Res. **83**, 83–104 (1995)
13. Psaraftis, H., Wilson, N., Jaw, J., Odoni, A.: A heuristic algorithm for the multi-vehicle many-to-many advance request dial-a-ride problem. Transp. Res. B **20B**, 243–257 (1986)
14. Madsen, O., Ravn, H., Rygaard, J.: A heuristic algorithm for the a dial-a-ride problem with time windows, multiple capacities, and multiple objectives. Ann. Oper. Res. **60**, 193–208 (1995)
15. Deleplanque, S., Quilliot, A.: Constraint Propagation for the Dial-a-Ride Problem with Split Loads. In: Recent Advances in Computational Optimization. Studies in Computational Intelligence, vol. 470, pp. 31–50. Springer, (2013)

# Multiple Shooting SQP Algorithm for Optimal Control of DAE Systems with Inconsistent Initial Conditions

Paweł Drąg and Krystyn Styczeń

**Abstract** In the article a new approach for control of differential-algebraic systems with inconsistent initial conditions were presented. The consistent initial conditions could be difficult to obtain, for example, in the fed-batch penicillin fermentation process. Additionally difficulties were incorporated by constraints on the differential state trajectories. For this purposes a new algorithm based on a multiple shooting SQP-line search method was proposed. To ensure a stability of the solution, the multiple shooting approach were used. By division a system into smaller subsystems, as a result a large-scale problem was obtained. The proposed algorithm can be applied to a wide class of differential-algebraic systems in booth chemical and mechanical engineering. The simulations were executed in Matlab environment using Wroclaw Centre for Networking and Supercomputing.

**Keywords** Optimal control · DAE systems · Multiple shooting method · State constraints · Inconsistent initial conditions

## 1 Introduction

Problems of efficient control of both complex purely dynamical and differential-algebraic systems are today an important task and appear commonly in a process design [1, 2, 15].

To provide a mathematical model of the process with slowly varying dynamics, often only algebraic equations are enough. But in some situations, when dynamics plays a key role in the system, differential equations have to be introduced to reflect the behavior of the system. Suitable numerical methods and optimization algorithms were proposed and implemented, so processes described by differential equations are

P. Drąg (✉) · K. Styczeń
Institute of Computer Engineering, Control and Robotics, Wrocław University of Technology, Janiszewskiego 11-17, 50-372 Wrocław, Poland
e-mail: pawel.drag@pwr.wroc.pl

K. Styczeń
e-mail: krystyn.styczen@pwr.wroc.pl

© Springer International Publishing Switzerland 2015                                                    53
S. Fidanova (ed.), *Recent Advances in Computational Optimization*,
Studies in Computational Intelligence 580, DOI 10.1007/978-3-319-12631-9_4

a group of well-known problems. An existence of a solution for all initial conditions, which is one of the most important feature of the differential systems, is applied for optimization of the multistage processes. Difficulties may, however, be caused by the instability of the mathematical model and selection of the appropriate numerical methods for the equations has to be made [9].

Usually, a real-life chemical processes are described by both algebraic and differential relations. In this manner, one can obtain the system of equations, which consists of a part of solely differential and algebraic equations. The mathematical model of this type is desirable, because during the construction of the model one does not need to perform additional transformations to obtain allowed equations. Other important feature is, that the variables in a model are known to have physical interpretation. When the equations are well scaled, then no other transformations are needed. Additionally, one can explore the impact of different variables on the behavior of the model [3].

The main topic of the article is the searching for the optimal control of the system governed by differential-algebraic equations. It is a challenge, because solution of the initial value problem for DAE systems do not exists for all possible values of parameters [5, 12, 13].

In this paper an optimization algorithm for the large-scale optimal control problem was presented. The large number of both optimization variables and constraints are introduced by the path constraints on the state trajectory [2]. In this manner a huge computational effort can be required by systems with simple path constraints, but with large number of decision variables.

The system of differential-algebraic equations was used to describe the fed-batch penicillin fermentation process [10]. Because there is a constraint on the state trajectory, the discretized nonlinear programming problem with infinitely many decision variables has to be considered. So, the practical approach leads to the use of the existing finite-dimensional methods adjusted to large scale optimization problems.

To solve the mentioned control problem, the new multi-step algorithm was designed, which can take into account the large number of variables. The study was carried out on the large-scale task of about 6,000 variables and 3,000 differential and 2,000 algebraic equations [7]. The algorithm combines the multiple shooting method and the simultaneous approach.

The aim of this paper is to present the SQP algorithm that improves an initial solution of a large-scale problem in a reasonable time.

The article is structured as follows. At the beginning the optimal control problem of DAE system was formulated. Then the simultaneous approach for the optimal control of differential algebraic systems and its relationship with the multiple shooting method is discussed. The Multi-step SQP-line search algorithm using the multiple shooting method is presented. The differential-algebraic model of the fed-batch penicillin fermentation process was described and solved by the designed algorithm. Finally, the results of the large-scale simulations, which were performed using Wroclaw Centre for Networking and Supercomputing, were discussed.

## 2 The Multiple Shooting Approach for Optimal Control of the Differential-Algebraic Systems

In the paper the following multiple shooting approach for optimal control problem of differential-algebraic systems is considered

$$\min_{p} \phi(p) = \sum_{l=1}^{N_T} \Phi(z^l(t_l), y^l(t_l), p^l), \tag{1}$$

subject to

$$z^{l-1}(t_{l-1}) = z_0^l = 0; \quad l = 2, \ldots, N_T, \tag{2}$$

$$z^{N_T}(t_{N_T}) - z_f = 0; \quad z^l(0) = z_0^l, \tag{3}$$

$$p_L^l \le p^l \le p_U^l, \tag{4}$$

$$y_L^l \le y^l(t_l) \le y_U^l, \tag{5}$$

$$z_L^l \le z^l(t_l) \le z_U^l; \quad l = 1, \ldots, N_T, \tag{6}$$

with the index-1 DAE system

$$\frac{dz^l(t)}{dt} = f^l(z^l(t), y^l(t), p^l); \quad z^l(t_{l-1}) = z_0^l, \tag{7}$$

$$g^l(z^l(t), y^l(t), p^l) + \mathcal{B}_l = 0; t \in [t_{l-1}, t_l]; \quad l = 1, \ldots, N_T. \tag{8}$$

In Eqs. (1–8) $z(t)$ denotes the differential state trajectory, $y(t)$ denotes the algebraic state trajectory and $\mathcal{B}_l$ is an inconsistency parameter. The control profile is represented as a parametrized function with coefficients that determine the optimal profile [16, 17]. The decision variables on DAE equations appear only in the time independent vector $p$.

The assumption on the invertability of $g(-, y(t), -)$ permits an implicit elimination of the algebraic variables $y(t) = y[z(t), p]$ [5]. While there are $N_T$ periods in DAE equations, the time dependent bounds and other path on the state variables are no longer considered. The algebraic constraints and terms in the objective function are applied only at the beginning of each period.

The optimal control problem of the fed batch penicillin fermentation process is an example of wide range control problems of systems described by nonlinear differential-algebraic equations (e.g. [15]), with difficult to obtain consistent initial conditions.

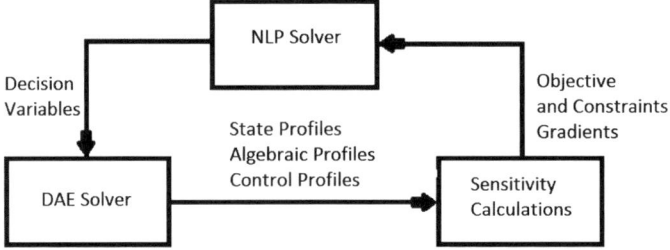

**Fig. 1** Sequential dynamic optimization strategy

Because of the instability, which can occur, shooting methods were developed. The shooting method was adjusted for solving more difficult systems and is usually known as the multiple shooting method or the parallel shooting method. When the multiple shooting approach is used, the time domain is partitioned into smaller time periods and the DAE models are integrated separately in each element. To provide the continuity of the states across elements, the equality constraints are added to the nonlinear program. The inequality constraints for states and controls are then imposed directly at the grid points $t_l$ [1].

The aim of the simultaneous approach is searching for the optimal control trajectory, the differential and algebraic state trajectories in a special manner. A sketch of the sequential dynamic optimization strategy for the problem (1–8) is presented on Fig. 1. At $l$-iteration, the variables $p^l$ are specified by NLP solver. In this situation, when the values of $p^l$ are known, one can treat DAE system as an initial value problem and integrate (2–4) forward in time for periods $l = 1, \ldots, N_T$. For these purposes Backward Differentiation Formula was used, which can solve index-1 DAEs. The differential state profile, the algebraic state profile and the control function profile were obtained as results of this step. Next component evaluates the gradient of the objective and constraint functions with respect to $p^l$. Because function and gradient information are passed to the NLP solver, then the decision variables can be updated [4, 8].

## 3 Dynamic Optimization of the Fed-Batch Penicillin Fermentation Process

The solution of a singular control problem is required to determine the optimal feeding profile of the fed-batch fermentation. Because there are nonlinear dynamics of the system model and constraints on both the state and control variables, the considered problem is difficult. The model used here was described and solved in [10]. The formulation is as follows

$$\min_{S(t)} J = \int_{t_0}^{t_f} \left( -\theta XV + 0.0103PV \right. \tag{9}$$

$$\left. + \ 0.0744\mu X + 0.00102XV + 6913.58 \right) dt$$

subject to

$$\dot{X} = \mu X - \frac{FX}{V}, \tag{10}$$

$$\dot{P} = \theta X - \frac{FP}{V}, \tag{11}$$

$$\dot{V} = F, \tag{12}$$

$$0 = \mu - \frac{0.11S}{S + 0.006X}, \tag{13}$$

$$0 = \theta - \frac{0.004}{1 + \frac{0.0001}{S} + \frac{S}{0.1}}, \tag{14}$$

$$0.001 \leq S \leq 0.5, \tag{15}$$

$$X \leq 41, \tag{16}$$

$$\begin{bmatrix} X(0), & P(0), & V(0) \end{bmatrix} = \begin{bmatrix} 1.0, & 0.0, & 2.5E5 \end{bmatrix}. \tag{17}$$

The overall profit is represented by the performance index (Eq. 9). There are three differential state variables:

(i) the concentration of the biomass $X(g/l)$,
(ii) the amount of existing penicillin product $P$ (activity per $l$),
(iii) the reactor volume $V(l)$ and two algebraic state variables in the system:
(iv) the specific growth rate $\mu$,
(v) the specific product formation rate $\theta$.

The control variable is the substrate concentration $S(g/l)$. The feed rate $F$ was set to 1666.67 $l/h$.

## 4 The Multi-step SQP Line Search Algorithm

The designed algorithm belongs to a group of the Sequential Quadratic Programming methods [14]. Its main part is as follows.

The equality constrained problem is considered

$$\min_p f(p),\qquad(18)$$

subject to

$$c(p) = 0,\qquad(19)$$

where the objective function $f : \mathcal{R}^n \to \mathcal{R}$ and the vector of equality constraints $c : \mathcal{R}^n \to \mathcal{R}^m$ are smooth functions. The idea behind the SQP approach is to model (18), (19) at the current iterate $p_k$ by a quadratic programming subproblem. Then the subproblem is minimized and the new iterate $p_{k+1}$ is defined.

The Lagrangian function for this problem is

$$\mathcal{L}(p, \lambda) = f(p) - \lambda^T c(p).\qquad(20)$$

The matrix $A(p)$ were used to denote the Jacobian matrix of the constraints

$$A(p) = [\nabla c_1(p), \nabla c_2(p), \dots, \nabla c_m(p)]^T,\qquad(21)$$

where $c_i(p)$ is the $i$th component of the vector $c(p)$.

The first order KKT conditions of the equality constrained problem (18), (19) can be written as the system on $n + m$ equations and the $n + m$ unknowns $p$ and $\lambda$,

$$\begin{bmatrix} \nabla f(p) - A(p)^T \lambda \\ c(p) \end{bmatrix} = 0.\qquad(22)$$

Any solution $(p^*, \lambda^*)$ of the equality constrained problem (18), (19) for which $A(p^*)$ has full row rank satisfies (22). The nonlinear system (22) can be solved by the Newton method.

The Jacobian of (22) with respect to $p$ and $\lambda$ is given by

$$\begin{bmatrix} \nabla^2_{pp}\mathcal{L}(p, \lambda) & -A(p)^T \\ A(p) & 0 \end{bmatrix} = 0.\qquad(23)$$

The Newton step from the iterate $(p_k, \lambda_k)$ is given by

$$\begin{bmatrix} p_{k+1} \\ \lambda_{k+1} \end{bmatrix} = \begin{bmatrix} p_k \\ \lambda_k \end{bmatrix} + \begin{bmatrix} d_k \\ d_\lambda \end{bmatrix},\qquad(24)$$

where $d_k$ and $d_\lambda$ solve the Newton-KKT system

$$\begin{bmatrix} \nabla^2_{pp}\mathcal{L}(p, \lambda) & -A(p)^T \\ A(p) & 0 \end{bmatrix} = \begin{bmatrix} d_k \\ d_\lambda \end{bmatrix} + \begin{bmatrix} -\nabla f(p) + A(p)^T \lambda \\ -c(p) \end{bmatrix}.\qquad(25)$$

The Newton step is well defined when KKT matrix in (23) is nonsingular. This is satisfied, when the following assumptions hold [14]

**Assumption 1** The Jacobian of the constraints $A(p)$ has full row rank.

**Assumption 2** The matrix $\nabla^2_{pp}\mathcal{L}(p, \lambda)$ is positive definite on the tangent space of the constraints, that is, $d^T \nabla^2_{pp}\mathcal{L}(p, \lambda)d > 0$ for all $d \neq 0$ such that $A(p)d = 0$.

Suppose that at the iterate $(p_k, \lambda_k)$ the problem (18), (19) is modeled by the quadratic program

$$\min_p \; f_k + \nabla f_k^T p + \frac{1}{2}\nabla^2_{pp}\mathcal{L}_k p, \tag{26}$$

subject to

$$A_k(p) + c_k = 0. \tag{27}$$

If Assumptions 1 and 2 hold, then this problem has the unique solution $(d_k, l_k)$ that satisfies

$$\nabla^2_{pp}\mathcal{L}_k d_k + \nabla f_k - A_k^T l_k = 0, \tag{28}$$

$$A_k d_k + c_k = 0. \tag{29}$$

The vectors $d_k$ and $l_k$ can be identified with the solution of the Newton equation (25).

---

ALGORITHM 1. Local SQP Algorithm for solving
the equality constrained problem

---

Choose an initial par $(p_0, \lambda_0)$;
    (if $p_0$ is given, then $\lambda_0$ is given by Eq. (22))
Set $k \leftarrow 0$;
REPEAT UNTIL convergence test is satisfied
    evaluate $f_k, \nabla f_k, \nabla^2_{pp}\mathcal{L}_k, c_k, A_k$;
    solve (26), (27) to obtain $d_k$ and $l_k$;
    set $p_{k+1} \leftarrow p_k + d_k$;
    set $\lambda_{k+1} \leftarrow l_k$;
END (REPEAT)

---

On this basis, the new algorithm was designed.

---

ALGORITHM 2. The line search SQP algorithm

---

choose parameters $\eta \in (0, 0.5)$, $\tau \in (0, 1)$
and an initial pair $(p_0, \lambda_0)$;
evaluate $f(p_0)$, $\nabla f(p_0)$, $c_i(p_0)$,
$A_0 = [\nabla c_1(p_0), \nabla c_2(p_0), \ldots, \nabla c_m(p_0)]^T$;
if a quasi-Newton approximation is used, choose
   an initial $n \times n$ symmetric positive definite Hessian
   approximation $B_0$, otherwise compute $\nabla^2_{pp}\mathcal{L}_0$;
WHILE convergence test is not satisfied DO
   compute $d_k$ by solving (25),    let $\lambda$ be the corresponding multiplier;
   $d_\lambda \leftarrow \hat{\lambda} - \lambda_k$;
   choose $\mu_k$ to satisfy Eq. (30) with $\sigma = 1$;
   set $\alpha_k \leftarrow 1$;
   WHILE $\Phi_1(p + \alpha_k d_k; \mu_k) >$
     $\Phi_1(p_k; \mu_k) + \eta \alpha_k D_1(f(p_k; \mu_k); d_k)$ DO
     reset $\alpha_k \leftarrow \tau_\alpha \alpha_k$ for some $\tau_\alpha \in (0, \tau]$;
   END (WHILE)
   set $p_{k+1} \leftarrow p_k + \alpha_k d_k$ and $\lambda_{k+1} \leftarrow \lambda_k + \alpha_\lambda d_\lambda$;
   IF a quasi-Newton approximation is used THEN
     set $s_k \leftarrow \alpha_k d_k$;
     set $\hat{y}_k \leftarrow \nabla_p \mathcal{L}(p_{k+1}, \lambda_{k+1}) - \nabla_p \mathcal{L}(p_k, \lambda_{k+1})$;
     obtain $B_{k+1}$ by updating $B_k$ using
     a quasi-Newton formula
     $B_{k+1} = B_k + \frac{(\hat{y}_k - B_k s_k)(\hat{y}_k - B_k s_k)^T}{(\hat{y}_k - B_k s_k)^T s_k}$
   END (IF)
END (WHILE)

---

The strategy for choosing $\mu$ in the Algorithm 2 considers the effect of the step on a model of the merit function, so $\mu$ has to satisfy the inequality

$$\mu \geq \frac{\nabla f_k^T d_k + \frac{\sigma}{2} d_k^T \nabla^2_{pp}\mathcal{L}_k d_k}{(1 - \rho)\|c_k\|_1}. \tag{30}$$

If the value of $\mu$ from the previous iteration of the SQP method satisfies Eq. (30), it is left unchanged. Otherwise, $\mu$ is increased, so that satisfies this inequality with some margin. The constant $\sigma$ is used to handle the case in which Hessian $\nabla^2_{pp}\mathcal{L}_k$ is not positive definite. We define $\sigma = 1$ if $d_k^T \nabla^2_{pp}\mathcal{L}_k d_k > 0$, and $\sigma = 0$ otherwise.

The $l_1$ merit function for the problem (18), (19) takes the form

$$\Phi_1(p; \mu) = f(p) + \mu\|c_k\|_1. \tag{31}$$

The directional derivative of $\Phi_1$ in the direction $d_k$ satisfies

$$D(\Phi_1(p_k; \mu); d_k) = \nabla f_k^T d_k - \mu\|c_k\|_1. \tag{32}$$

---

ALGORITHM 3. The SQP-line search algorithm
for solving the equality constrained problem

---

BEGIN
   define a vector of decision variables $\tilde{p}$
     and its initial conditions;
   choose from vector $\tilde{p}$ a subvector $p$,
     which describes a subsystem
     $\mathcal{S} = f(p)$
   solve problem (36) using Algorithm 2;
   update values of vector $\tilde{p}$ using results
     from the previous step;
END

---

As one can see, the Algorithm 2 can be thought as an inner loop in the Algorithm 3. The last question is, what is the rate of convergence of the considered algorithm.

**Assumption 3** The point $p^*$ is a local solution of the problem (18), (19) at which the following conditions hold.

a) The functions $f$ and $c$ are twice differentiable in a neighborhood of $p^*$ with Lipschitz continuous second derivatives.

b) The linear independence constraint qualification holds at $p^*$.

c) The second order sufficient conditions hold at $(p^*, \lambda^*)$.

Now one can call the theorem, which justifies the correctness of the presented algorithm.

**Theorem 4** *[14]: Suppose, that Assumption 3 holds and that the iterates $p_k$ generated by Algorithm 1 with quasi-Newton approximate Hessian $B_k$, converge to $p^*$. Then $p_k$ converges superlinearly if and only if the Hessian approximation satisfies*

$$\lim_{k \to \infty} \frac{\|(B_k - \nabla_{pp}^2 \mathcal{L}_*)(p_{k+1} - p_k)\|}{\|p_{k+1} - p_k\|} = 0. \tag{33}$$

**Lemma 5** *Algorithm 3 generates a sequence of the feasible solutions with decreasing values of the goal function. In this bounded sequence one can distinguish a subsequence, which is superlinearly convergent to the locally optimal solution $p^*$.*

## 5 Numerical Results

Simulations were executed on the large-scale fed-batch penicillin fermentation model.

At the beginning, the reactor was divided into 20 parts and the solution was obtained in 5 h. At this step the vector of decision variables was stated as follows

$$p = [u_1, \ldots, u_{20}, X_{0,2}, \ldots, X_{0,20}, P_{0,2}, \ldots, P_{0,20}, \tag{34}$$
$$V_{0,2}, \ldots, V_{0,20}, \mu_{0,2}, \ldots, \mu_{0,20}, \theta_{0,2}, \ldots, \theta_{0,20}].$$

Solution of this model was used as the initial conditions in the further work.

The question is, how to choose the vector of decision variables, to obtain in a reasonable time a possibly greatest improvement of the solution.

Then the reactor was divide into 1,000 parts. There are 5,997 decision variables in the system (1,000 piecewise constant control functions and 4,997 variables treated as initial conditions for differential and algebraic state trajectories).

$$\widetilde{p} = [u_1, \ldots, u_{1000}, X_{0,2}, \ldots, X_{0,1000}, P_{0,2}, \ldots, P_{0,1000}, \tag{35}$$
$$V_{0,2}, \ldots, V_{0,1000}, \mu_{0,2}, \ldots, \mu_{0,1000}, \theta_{0,2}, \ldots, \theta_{0,1000}].$$

The simulations were executed for four different possible number of variables in the subvector $\widetilde{p}$: 10, 20, 50 and 100. As decision variable control function and initial conditions of DAE systems, especially in the initial phases of the process, were considered. This enables increase the accuracy of the calculation, when the reactions proceed quickly. The initial average nonconsistency of algebraic equations in DAE models was $\bar{d} = 1.0816e - 5$ and $J = -1.3691e6$.

In the simulations two different stop criteria in the Algorithm 2 were used. In the implementation convergence was declared when $TolFun < \epsilon_1$ and $TolCon < \epsilon_2$. $TolFun$ denotes termination tolerance on the function value, and $TolCon$ denotes tolerance on the constraint violation.

**Cases 1** In the first case the local optimization processes were performed more precisely. So, $TolFun < 1e - 6$ and $TolCon < 1e - 6$.

**Cases 2** In the second case the stop criterion in the local optimization was no so rigorous: $TolFun < 1e - 3$ and $TolCon < 1e - 3$.

The main stop criterion was the performance time. When the computing time exceeded 5 h, optimization process was stopped.

In both cases the augmented objective function was considered

$$f(p) = J + \rho \sum_{l=1}^{N_T} (z_0^{l+1} - \hat{z}_l)^2 + \rho \sum_{l=1}^{N} \mathcal{B}_l^2, \tag{36}$$

where penalty parameter $\rho = 10^7$ and $\mathcal{B}_l(z^l(t_l), y^l(t_l), p^l)$ denotes the inconsistency of the initial conditions.

Equation (36) shows the balance in the quest to minimize the function $J$ and to meet both the continuity constraints and consistency of the initial conditions in differential-algebraic equations.

Results presented in the Tables 1 and 2 show, that the inexact algorithm with the weak stop criteria, can obtain a better improvement of the optimization problem.

**Table 1** Results of the simulations in case 1

| Size of subvector $p$ | Number of iterations | $\bar{d}$ | $J$ |
|---|---|---|---|
| 10 | 67 | $0.5344e - 3$ | $-1.4738e6$ |
| 20 | 32 | $0.5259e - 3$ | $-1.4437e6$ |
| 50 | 15 | $0.5613e - 3$ | $-1.6651e6$ |
| 100 | 6 | $0.5145e - 3$ | $-1.4196e6$ |

**Table 2** Results of the simulations in case 2

| Size of subvector $p$ | Number of iterations | $\bar{d}$ | $J$ |
|---|---|---|---|
| 10 | 73 | $0.5538e - 3$ | $-1.5967e6$ |
| 20 | 41 | $0.5858e - 3$ | $-1.6982e6$ |
| 50 | 15 | $0.5612e - 3$ | $-1.6649e6$ |
| 100 | 6 | $0.5145e - 3$ | $-1.4196e6$ |

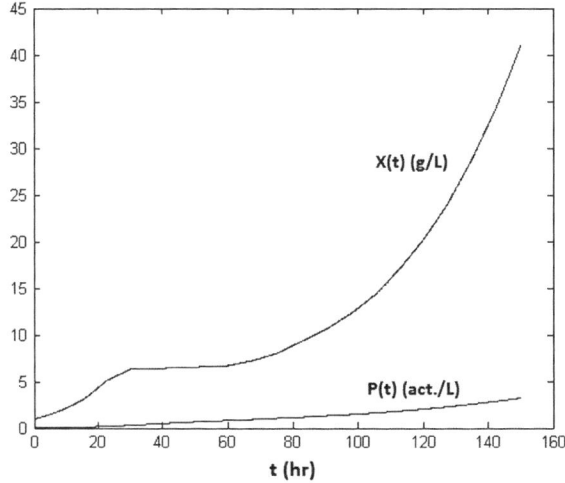

**Fig. 2** The optimal state trajectories. Size of the subvector $p = 20$ and stop criteria like in the case 2

As a result, the final concentration was improved and the obtained solution meets the consistency conditions with high accuracy.

The solutions obtained for size of the subvector $p = 20$ and stop criteria like in case 2 were presented on the Figs. 2 and 3. There are the differential state trajectories in the Fig. 2 and the optimal control profile in the Fig. 3.

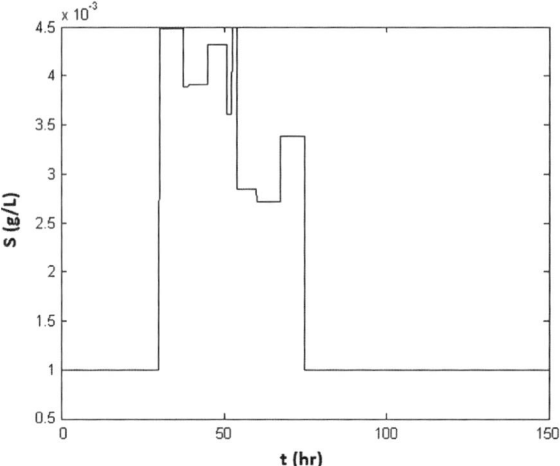

**Fig. 3** The optimal control profile. Size of the subvector $p = 20$ and stop criteria like in the case 2

## 6 Conclusion

In the article the control of the fed-batch penicillin fermentation was considered. The complex model of the reactor was designed using the direct multiple shooting approach. For this purposes the new SQP-line search algorithm was designed and tested. The algorithm, which takes in each iteration only a few number of decision variables into account, can do new iterations and improve the initial solution. But in both approaches a large number of variables were considered.

In its pure form, SQP algorithm is convergent to the locally solution. The line search was used as a globalization approach to construct a sequence of feasible solutions with decreasing values of the objective function.

This type of algorithms can be successfully applied to the large systems, when structure of Hessian matrix can not be effectively used. It is worth to note, that presented procedure is a modified algorithm presented in [6], which was applied to control of the pressure-constrained batch reactor.

Second order information, which can be approximated using BFGS method, can be unavailable when Jacobian matrix is difficult to calculate. This situation one can be met very often, when simultaneous approach is used.

The multistep algorithms, which need consistent and feasible initial conditions, can improve the solution in considerable short time. At the end we want to emphasize the need for booth Jacobian-free optimization algorithms, which could solve the large-scale optimization tasks [11] and solvers for the real-life optimization and optimal control problems.

**Acknowledgments** The project was supported by the grant of National Science Centre Poland DEC-2012/07/B/ST7/01216.

# References

1. Betts, J.T.: Practical methods for optimal control and estimation using nonlinear programming, 2nd edn. SIAM, Philadelphia (2010)
2. Biegler, L.T.: Nonlinear programming. Concepts, algorithms, and applications to chemical processes. SIAM, Philadelphia (2010)
3. Biegler, L.T., Campbell, S.L., Mehrmann, V.: Control and optimization with differential-algebraic constraints. SIAM, Philadelphia (2012)
4. Biegler, L.T., Grossmann, I.E.: Retrospective on optimization. Comput. Chem. Eng. **28**, 1169–1192 (2004)
5. Brenan, K.E., Campbell, S.L., Petzold, L.R.: Numerical solution of initial-value problems in differential-algebraic equations. SIAM, Philadelphia (1996)
6. Drąg, P., Styczeń, K.: Multiple shooting SQP-line search algorithm for optimal control of pressure-constrained batch reactor. In: Proceedings of the Federated Conference on Computer Science and Information Systems, pp. 307–313 (2013)
7. Feehery, W.E., Tolsma, J.E., Barton, P.I.: Efficient sensitivity analysis of large-scale differential-algebraic systems. Appl. Numer. Math. **25**, 41–54 (1997)
8. Grossmann, I.E., Biegler, L.T.: Part II. Future perspective on optimization. Comput. Chem. Eng. **28**, 1193–1218 (2004)
9. Hairer, E., Lubich, C., Roche, M.: The numerical solution of differential-algebraic systems by runge-kutta method. Lecture Notes in Mathematics. Springer, Berlin (1989)
10. Huang, Y.J., Reklaitis, G.V., Venkatasubramanian, V.: Model decomposition based method for solving general dynamic optimization problems. Comput. Chem. Eng. **26**, 863–873 (2002)
11. Knoll, D.A., Keyes, D.E.: Jacobian-free Newton-Krylov methods: a survey of approaches and applications. J. Comput. Phys. **193**, 357–397 (2004)
12. Leineweber, D.B., Schaefer, A., Bock, H.G., Schloeder, J.P.: An efficient multiple shooting based reduced SQP strategy for large-scale dynamic process optimization. Part II: Software aspects and applications. Comput. Chem. Eng. **27**, 167–174 (2003)
13. Maerz, R.: Numerical methods for differential algebraic equations. Acta Numer. 141–198 (1992)
14. Nocedal, J., Wright, S.J.: Numerical Optimization, 2nd edn. Springer, New York (2006)
15. von Schwerin, M., Deutschmann, O., Schulz, V.: Process optimization of reactive systems by partially reduced SQP methods. Comput. Chem. Eng. **24**, 89–97 (2000)
16. Vassiliadis, V.S., Sargent, R.W.H., Pantelides, C.C.: Solution of a class of multistage dynamic optimization problems. 1. Problems without path constraints. Ind. Eng. Chem. Res **33**, 2111–2122 (1994)
17. Vassiliadis, V.S., Sargent, R.W.H., Pantelides, C.C.: Solution of a class of multistage dynamic optimization problems. 2. Problems with path constraints. Ind. Eng. Chem. Res **33**, 2123–2133 (1994)

# A Constructive Algorithm for Partial Latin Square Extension Problem that Solves Hardest Instances Effectively

Kazuya Haraguchi

**Abstract**  A partial Latin square (PLS) is a partial assignment of $n$ symbols to an $n \times n$ grid such that, in each row and in each column, each symbol appears at most once. The partial Latin square extension (PLSE) problem asks to find such a PLS that is a maximum extension of a given PLS. Recently Haraguchi et al. proposed a heuristic algorithm for the PLSE problem. In this paper, we present its effectiveness especially for the "hardest" instances. We show by empirical studies that, when $n$ is large to some extent, the instances such that symbols are given in 60–70 % of the $n^2$ cells are the hardest. For such instances, the algorithm delivers a better solution quickly than IBM ILOG CPLEX, a state-of-the-art optimization solver, that is given a longer time limit. It also outperforms surrogate constraint based heuristics that are originally developed for the maximum independent set problem.

**Keywords**  Partial Latin square extension problem · Heuristic · Constructive algorithm · Maximum matching · Maximum independent set problem

## 1 Introduction

Throughout the paper, we consider the *partial Latin square extension* (*PLSE*) problem. Let $n \geq 2$ be a natural number. Suppose that we are given an $n \times n$ grid of *cells*. A *partial Latin square* (*PLS*) is a partial assignment of $n$ *symbols* to the grid so that the *Latin square condition* is satisfied. The Latin square condition requires that, in each row and in each column, every symbol should appear at most once. Given a PLS, the PLSE problem asks to construct such a PLS that is a maximum extension of the given one in terms of the number of the filled cells. Specifically, we are asked to assign symbols to as many empty cells as possible as long as the Latin square condition is satisfied. The PLSE problem is NP-hard since its decision problem version

This work is partially supported by JSPS KAKENHI Grant Number 25870661.

K. Haraguchi (✉)
Faculty of Commerce, Otaru University of Commerce, Otaru, Japan
e-mail: haraguchi@res.otaru-uc.ac.jp

© Springer International Publishing Switzerland 2015
S. Fidanova (ed.), *Recent Advances in Computational Optimization*,
Studies in Computational Intelligence 580, DOI 10.1007/978-3-319-12631-9_5

is NP-complete [1]. The problem has various applications (e.g., scheduling, optical routers, combinatorial puzzles [2, 3]) and was first introduced as an optimization problem by Kumar et al. [4].

In this paper, we extend the result of the preliminary paper which was written by Haraguchi et al. [5]. They proposed a heuristic algorithm for the PLSE problem that utilizes the notion of maximum matching. Hereafter let us call this algorithm MATCHING. They observed how the performance of MATCHING improves or deteriorates as they change the internal settings, and how it is effective for large-scale "hard" instances in comparison with IBM ILOG CPLEX (CPLEX for short) [6], a state-of-the-art optimization solver.

This paper has two highlights. (I) We make detailed investigation into which instances are the "hardest." In the preliminary paper, the notion of the instance hardness was just based on the well-known easy-hard-easy phase transition. Hence the specialty of MATCHING was rather vague. Conducting intensive studies, we find that, when $n$ is large to some extent, the instances such that symbols are given in 60 % to 70 % of the $n^2$ cells are the hardest in a certain sense. (II) We present how MATCHING is effective for such hardest instances. We compare it with CPLEX and surrogate constraint based heuristics [7, 8] in terms of performance on the hardest instances. The PLSE problem can be formulated as a *maximum independent set* (*MIS*) problem. Surrogate constraint based heuristics are originally developed for the MIS problem, and we apply them to the PLSE problem in MIS formulation as benchmark heuristics.

The paper is organized as follows. In Sect. 2, we introduce notations and terminologies. We review the algorithm MATCHING in Sect. 3. We report the experimental studies on which instances are the "hardest" in Sect. 4, and then in Sect. 5, we make performance comparison between MATCHING and other algorithms. Finally we give concluding remarks in Sect. 6.

# 2 Preliminaries

## *2.1 Graph, Maximum Matching, Independent Set*

An *undirected graph* (or simply a *graph*) $G = (V, E)$ consists of a set $V$ of *nodes* and a set $E$ of unordered pairs of nodes, where each element in $E$ is called an *edge*. When two nodes are joined by an edge, we say that they are *adjacent*, or equivalently, that one node is a *neighbor* of the other. The *degree* of a node $v \in V$ is the number of its neighbors. A graph is *bipartite* when $V$ can be partitioned into two disjoint nonempty sets, say $V_1$ and $V_2$, so that every edge joins a node in $V_1$ and a node in $V_2$. When we emphasize that $G$ should be bipartite, we may write $G = (V_1 \cup V_2, E)$.

A *matching* $E'$ is a subset of $E$ such that no two edges in $E'$ have a node in common. A *maximum matching* is such a matching that attains the maximum cardinality. The size of maximum matching is called a *matching number*, and is denoted by $\nu(G)$.

Suppose that the graph $G = (V, E)$ is bipartite. We can find a maximum matching in $O(\sqrt{|V|}|E|)$ time [9]. Note that maximum matching is not necessarily unique. Any edge $e$ in $E$ is classified into one of the following three classes with respect to how it appears in the possible maximum matchings:

**Mandatory edge:**   $e$ appears in *every* maximum matching.
**Admissible edge:**   $e$ appears in *at least one* (but not every) maximum matching.
**Forbidden edge:**   $e$ appears in *no* maximum matching.

The sets of mandatory, admissible and forbidden edges in $G$ are denoted by $ME(G)$, $AE(G)$ and $FE(G)$, respectively. The edge set $E$ of a bipartite graph $G$ can be decomposed into three disjoint sets $ME(G)$, $AE(G)$ and $FE(G)$ by the Dulmage-Mendelsohn decomposition technique [10]. The computation time is dominated by finding a maximum matching, and thus the decomposition can be made in $O(\sqrt{|V|}|E|)$ time. The algorithm MATCHING repeatedly solves maximum matching problems on bipartite graphs such that $|V_1| = |V_2| = n$. We denote the upper bound on the computation time for one bipartite graph by $\tau_n$, i.e., $\tau_n = O(n^{5/2})$.

Let $G = (V, E)$ be a general graph, not necessarily bipartite. An *independent set* is a subset $V' \subseteq V$ of nodes such that no two nodes in $V'$ are adjacent. The maximum independent set (MIS) problem asks to find an independent set of the maximum cardinality for a given $G$. It is a well-known NP-hard problem [11].

## 2.2 PLSE Problem

First we formulate the PLSE problem in a rather conventional way. We then give its formulation by integer programming (IP), by constraint programming (CP) and by MIS problem.

Let $n \geq 2$ be a positive integer. We denote $[n] = \{1, 2, \ldots, n\}$. Suppose an $n \times n$ grid of cells. For any $i, j \in [n]$, we denote the cell in the row $i$ and in the column $j$ by $(i, j)$. We consider a partial assignment of $n$ symbols to the grid. The $n$ symbols to be assigned are the $n$ integers in $[n]$. A partial assignment of symbols is represented by a partial function $\varphi : [n] \times [n] \to [n]$, where $\varphi(i, j)$ indicates the symbol assigned to $(i, j)$. A cell $(i, j)$ is *empty* when $\varphi(i, j)$ is not defined. We denote the domain of $\varphi$ by $\mathrm{Dom}(\varphi) = \{(i, j) \in [n] \times [n] \mid \varphi(i, j) \text{ is defined}\}$. In other words, $\mathrm{Dom}(\varphi)$ is the set of non-empty cells. A partial assignment $\varphi'$ is an *extension* of $\varphi$ when $\mathrm{Dom}(\varphi') \supseteq \mathrm{Dom}(\varphi)$ and $\varphi'(i, j) = \varphi(i, j)$ for every $(i, j) \in \mathrm{Dom}(\varphi)$. We call $\varphi$ a *partial Latin square (PLS) function* if it satisfies the following Condition 1.

**Condition 1 (Latin square condition in partial function representation).** For any $i = 1, \ldots, n$, every symbol in $[n]$ appears in $\varphi(i, 1), \ldots, \varphi(i, n)$ at most once. Similarly, for any $j = 1, \ldots, n$, every symbol in $[n]$ appears in $\varphi(1, j), \ldots, \varphi(n, j)$ at most once.

In particular, we call $\varphi$ simply a *Latin square (LS)* if it is total (i.e., $\mathrm{Dom}(\varphi) = [n]^2$). Given a PLS function $\varphi$, the partial Latin square extension (PLSE) problem asks to construct an extension $\varphi'$ of $\varphi$ so that $\varphi'$ is a PLS function and $|\mathrm{Dom}(\varphi')|$ attains the maximum.

**IP Formulation.** The PLSE problem can be formulated as a 0–1 integer programming problem. Let us introduce a binary variable $x_{ijk}$ for any $i, j, k \in [n]$. The $x_{ijk}$ indicates whether the symbol $k$ is assigned to the cell $(i, j)$, i.e.,

$$x_{ijk} = \begin{cases} 1 & \text{if } (i, j) \in \mathrm{Dom}(\varphi) \text{ and } \varphi(i, j) = k, \\ 0 & \text{otherwise.} \end{cases}$$

The PLSE problem is then formulated as a 0–1 integer programming problem as follows.

$$\text{maximize} \quad \sum_{(i,j,k)\in[n]^3} x_{ijk}$$

$$\text{subject to} \quad \forall(j, k) \in [n]^2, \sum_{i\in[n]} x_{ijk} \leq 1,$$

$$\forall(i, k) \in [n]^2, \sum_{j\in[n]} x_{ijk} \leq 1,$$

$$\forall(i, j) \in [n]^2, \sum_{k\in[n]} x_{ijk} \leq 1,$$

$$\forall(i, j) \in \mathrm{Dom}(\varphi), \ x_{ij}x_{ij\varphi(i,j)} = 1,$$

$$\forall(i, j, k) \in [n]^3, \ x_{ijk} \in \{0, 1\}.$$

**CP Formulation.** To formulate the PLSE problem as a constraint optimization problem, let us introduce an integer variable $y_{ij} \in [n] \cup \{0\}$ for any cell $(i, j) \in [n]^2$ on the grid; if $y_{ij} = 0$, then it is indicated that $(i, j)$ is empty, and otherwise, it is indicated that $(i, j)$ is assigned the integer $y_{ij} > 0$. Then the formulation is as follows.

$$\text{maximize} \quad |\{(i, j) \mid y_{ij} \neq 0\}|$$

$$\text{subject to} \quad \forall i \in [n], \ \texttt{all-different\_except\_0}(y_{i1}, \ldots, y_{in}),$$

$$\forall j \in [n], \ \texttt{all-different\_except\_0}(y_{1j}, \ldots, y_{nj}),$$

$$\forall(i, j) \in \mathrm{Dom}(\varphi), \ y_{ij} = \varphi(i, j),$$

$$\forall(i, j) \in [n]^2, \ y_{ij} \in [n] \cup \{0\}.$$

In the above formulation, the `all-different_except_0` constraint is an extension of the typical all-different constraint, requiring that the variables should take all-different values except the variables assigned 0.

**MIS Formulation.** We need some preparations before we formulate the PLSE problem by the MIS problem. Let us represent a partial assignment by a set of triples, say

$T \subseteq [n]^3$. The membership $(v_1, v_2, v_3) \in T$ represents that the symbol $v_3$ is assigned to $(v_1, v_2)$. To avoid a duplicate assignment, we assume that, for any two triples $v = (v_1, v_2, v_3)$ and $w = (w_1, w_2, w_3)$ in $T$ $(v \neq w)$, $(v_1, v_2) \neq (w_1, w_2)$ holds. For any two triples $v, w \in [n]^3$, the Hamming distance between $v$ and $w$ is defined by the number of dimensions such that the two triples take different values. Denoted by $d(v, w)$, the Hamming distance is defined as $d(v, w) = |\{k \in [3] \mid v_k \neq w_k\}|$. In the sequel, we simply call $d(v, w)$ the distance between $v$ and $w$, omitting the term "Hamming." Obviously we have $d(v, w) \in \{0, 1, 2, 3\}$, where it is 0 if and only if $v$ and $w$ are equivalent.

It is easy to translate a partial function into the triple set; a partial function $\varphi :$ $[n] \times [n] \to [n]$ is translated into a triple set $T_\varphi = \{(i, j, \varphi(i, j)) \mid (i, j) \in [n] \times [n], \varphi(i, j) \text{ is defined}\}$. Condition 1 for partial function representation is translated into the following Condition 2. Specifically, $\varphi$ satisfies Condition 1 if and only if $T_\varphi$ satisfies Condition 2. Thus we call a triple set $T$ a *PLS set* if it satisfies Condition 2.

**Condition 2 (Latin square condition in triple set representation).** For any two triples $v, w \in T$ $(v \neq w)$, $d(v, w)$ is either 2 or 3.

**Proposition 1** *A partial function $\varphi$ satisfies Condition 1 if and only if the triple set $T_\varphi$ satisfies Condition 2.*

*Proof* We show the necessity by contradiction. Suppose that $\varphi$ satisfies Condition 1, but that $T_\varphi$ does not satisfy Condition 2. Then there exist $v, w \in T_\varphi$ $(v \neq w)$ such that $d(v, w) = 1$. We are in one of the following three cases: $(v_1 \neq w_1, v_2 = w_2$ and $v_3 = w_3)$, $(v_1 = w_1, v_2 \neq w_2$ and $v_3 = w_3)$, and $(v_1 = w_1, v_2 = w_2$ and $v_3 \neq w_3)$. The first two cases indicate that $\varphi(v_1, v_2) = \varphi(w_1, v_2) = v_3$ and $\varphi(v_1, v_2) = \varphi(v_1, w_2) = v_3$ respectively, both of which contradict that $\varphi$ satisfies Condition 1. The last case indicates that $\varphi$ assigns two symbols $v_3, w_3$ to $(v_1, v_2)$, contradicting that $\varphi$ is a function. We have seen that all the three cases never occur. The sufficiency can be shown in the similar manner. □

Let us denote a PLS set by $S \subseteq [n]^3$. Clearly any subset $S' \subseteq S$ is also a PLS set. We say that two PLS sets $S'$ and $S''$ are *compatible* if, for any $v \in S$ and $v' \in S'$, the distance $d(v, v')$ is either 2 or 3. The union of such $S'$ and $S''$ is a PLS, or equivalently, any PLS set $S$ is split into two disjoint PLS sets $S = S' \uplus S''$ such that $S'$ and $S''$ are compatible. The PLSE problem in triple set representation is formulated as follows; given a PLS set $L \subseteq [n]^3$, we are asked to find a PLS set of the maximum cardinality among the supersets of $L$. In other words, we are asked to construct a PLS set $S$ of the maximum cardinality such that $S$ and $L$ are compatible.

Now we are ready to transform a PLSE instance into an MIS instance. Suppose that we are given a PLSE instance in terms of a PLS set $L \subseteq [n]^3$. For any triple $v \in L$, we denote by $N^*(v)$ the set of all triples $w$'s such that $d(v, w) = 1$, i.e., $N^*(v) = \{w \in [n]^3 \mid d(v, w) = 1\}$. Clearly we have $|N^*(v)| = 3(n - 1)$. The union $\bigcup_{v \in L} N^*(v)$ over $L$ is denoted by $N^*(L)$. Then an MIS instance given by $G_L = (V_L, E_L)$ is equivalent to the PLSE instance $L$, where $V_L$ and $E_L$ are defined as follows:

$$V_L = [n]^3 \setminus (L \cup N^*(L)),$$
$$E_L = \{(v, w) \in V_L \times V_L \mid d(v, w) = 1\}. \tag{1}$$

**Proposition 2** *Let $S \subseteq [n]^3$ be a set of triples. Then $S$ is a feasible solution to the PLSE instance $L$ if and only if $S$, as a node set, is a feasible solution to the MIS instance $G_L = (V_L, E_L)$.*

*Proof* Let $S$ be a feasible solution to the PLSE instance $L$, that is, $S$ is a PLS set and compatible with $L$. From the compatibility, $S$ contains no triple in $L$ or in $N^*(L)$. Thus we have $S \subseteq [n]^3 \setminus (L \cup N^*(L)) = V_L$. Since $S$ is a PLS set, for any two triples $v, v' \in S$, the distance between $v$ and $v'$ is either 2 or 3. Then they are not joined by any edge in $E_L$. We see that $S$ is an independent set in $G_L$.

Let $S$ be an independent set in the MIS instance $G_L$. We have $S \subseteq V_L$ and no two triples $v, v' \in S$ are joined by an edge. The latter indicates that $d(v, v')$ is either 2 or 3 and thus that $S$ is a PLS set. The former indicates that $S$ is compatible with $L$.  □

By Proposition 2, we hereafter consider solving the PLSE instance by means of solving the transformed MIS instance $G_L = (V_L, E_L)$. Let us observe the structure of $G_L$ and introduce related notations. Recall that each node $v$ in $G_L$ is a triple $v = (v_1, v_2, v_3) \in [n]^3$. We regard $v$ as a grid point in the 3D integral space. Any grid point is an intersection of three grid lines that are orthogonal to each other. In other words, each node is on exactly three grid lines. Two nodes are joined by an edge if and only if they are on the same grid line. The nodes on the same grid line form a clique. We denote the set of $v$'-s neighbors by $N(v)$. The degree of $v$ is defined by $|N(v)|$. Obviously we have $|N(v)| \leq 3(n-1)$. Then since $|V_L| = O(n^3)$, we have $|E_L| = O(n^4)$.

## 3 Algorithm MATCHING

This section reviews the algorithm MATCHING that Haraguchi et al. proposed in the preliminary paper [5], where we use the MIS notation.

### 3.1 Generic Algorithm

First we describe the generic heuristic algorithm which MATCHING is based on. Starting from an empty set $S = \emptyset$, the generic algorithm constructs a final independent set of $G_L$ by repeating the following process: it chooses a certain subgraph of $G_{L \cup S}$, computes a maximum independent set $I$ of the subgraph, and updates $S$ by $S \leftarrow S \cup I$.

A subgraph that the algorithm chooses is induced by the nodes on the same 2D facet of $G_{L \cup S}$. Recall that $G_{L \cup S}$ exists in the 3D integral space. Among the 2D facets, the algorithm treats such ones that can be represented by a 2D plane $v_k = \ell$

**Algorithm 1** Generic algorithm for the PLSE problem

1: $S \leftarrow \emptyset$, $P = [3] \times [n]$
2: **while** $P$ is not empty **do**
3:     choose a pair $(k, \ell) \in P$
4:     $I \leftarrow$ a maximum independent set in $G_{LUS}^{k,\ell}$
                                $\triangleright$ $I$ can be constructed from a maximum matching in $B_{LUS}^{k,\ell}$.
5:     $S \leftarrow S \cup I$, $P \leftarrow P \setminus \{(k, \ell)\}$
6: **end while**

$(k \in [3], \ell \in [n])$. Then we have at most $3n$ facets and thus $3n$ subgraphs. Note that a facet (and thus a subgraph) is specified by a pair $(k, \ell) \in [3] \times [n]$. The set of the nodes on the 2D facet $v_k = \ell$ is defined as follows:

$$V_{LUS}^{k,\ell} = \{(v_1, v_2, v_3) \in V_{LUS} \mid v_k = \ell\}.$$

We denote by $G_{LUS}^{k,\ell}$ the subgraph induced by $V_{LUS}^{k,\ell}$.

A maximum independent set of $G_{LUS}^{k,\ell}$ can be computed in polynomial time. We describe the reason as follows. Let $\kappa, \kappa' \in [3]$ be dimension indices such that $\kappa < \kappa'$, $\kappa \neq k$ and $\kappa' \neq k$. We define the bipartite subgraph $B_{LUS}^{k,\ell} = (U \cup U', F_{LUS}^{k,\ell})$ such that $U = \{u_1, \ldots, u_n\}$ and $U' = \{u'_1, \ldots, u'_n\}$ are the node sets and $F_{LUS}^{k,\ell}$ is the edge set such that:

$$F_{LUS}^{k,\ell} = \{(u_i, u'_j) \in U \times U' \mid (v_1, v_2, v_3) \in V_{LUS}^{k,\ell}, \ v_\kappa = i, \ v_{\kappa'} = j\}.$$

From the definition, we see one-to-one correspondence between a node in $V_{LUS}^{k,\ell}$ and an edge in $F_{LUS}^{k,\ell}$. This leads to one-to-one, size-preserving correspondence between an independent set in $G_{LUS}^{k,\ell}$ and a matching in $B_{LUS}^{k,\ell}$; Suppose that an independent set $I$ of $G_{LUS}^{k,\ell}$ is given. Any two nodes $(v_1, v_2, v_3)$ and $(w_1, w_2, w_3)$ in $I$ should satisfy $v_\kappa \neq w_\kappa$ and $v_{\kappa'} \neq w_{\kappa'}$ since otherwise they are adjacent. Then the edge set $\{(u_i, u'_j) \mid (v_1, v_2, v_3) \in I, \ v_\kappa = i, \ v_{\kappa'} = j\}$ is a matching of $B_{LUS}^{k,\ell}$ of size $|I|$. For the converse direction, given a matching $M$ of $B_{LUS}^{k,\ell}$, the node set $\{(v_1, v_2, v_3) \in V_{LUS}^{k,\ell} \mid (u_i, u'_j) \in M, \ v_\kappa = i, \ v_{\kappa'} = j\}$ is an independent set of $G_{LUS}^{k,\ell}$ of size $|M|$. Computing a maximum matching $M$ of $B_{LUS}^{k,\ell}$, we obtain a maximum independent set of $G_{LUS}^{k,\ell}$ by converting $M$ into a node set as above.

Choosing $k \in [3]$ and $\ell \in [n]$ iteratively, the generic algorithm constructs a maximum independent set of $G_{LUS}^{k,\ell}$ (by means of computing a maximum matching of $B_{LUS}^{k,\ell}$) and appending it to $S$. The number of iteration times is at most $3n$ since we need not to choose each pair $(k, \ell)$ twice; once $(k, \ell)$ has been chosen to append a maximum independent set of $G_{LUS}^{k,\ell}$ to $S$, no nodes in $G_{LUS}^{k,\ell}$ can be appended to $S$ any more. This is because, when an independent set $I$ is appended to $S$, the nodes in $I \cup (\bigcup_{v \in I} N(v))$ and all the incident edges do not appear in the graph $G_{LUS}$. We summarize the generic algorithm in Algorithm 1. The set $P$ is introduced to manage which pair $(k, \ell)$ has been chosen or not.

The generic algorithm was inspired by the approximation algorithm discussed by Kumar et al. [4]. They analyzed a restricted version of the algorithm such that $P$ is initially set to $P = \{(c, 1), (c, 2), \ldots, (c, n)\}$ for a constant $c \in [3]$ in Line 1. They showed it to be a $1/2$-approximation algorithm, i.e., it is guaranteed that the algorithm delivers a solution whose size is at least half of the size of the global optimal solution.

## 3.2 Algorithm MATCHING

The algorithm MATCHING is a sophisticated version of Algorithm 1 in the following two points:

 (i) In Line 3, the pair $(k, \ell) \in P$ is chosen by means of the greedy method on a certain evaluating function $f : P \to \mathbb{Z}$.
 (ii) Before conducting Line 3, a set of "promising" nodes is added to the solution.

For (i), we employ the following evaluating function:

$$f_{L \cup S}(k, \ell) = v(B_{L \cup S}^{k,\ell}) + |\{(v_1, v_2, v_3) \in S \mid v_k = \ell\}|. \tag{2}$$

We choose the pair $(k, \ell) \in P$ that achieves the maximum $f_{L \cup S}(k, \ell)$. It must prefer $(k, \ell)$ such that more nodes can be newly added to $S$ (i.e., first term in (2)) and/or more nodes on the facet are already chosen in $S$ (i.e., second term in (2)). The only matching number in the first term appears to be a more natural criterion, but it did not yield good results in our preliminary experiments.

Further, when there are plural $(k, \ell)$'s such that $f_{L \cup S}(k, \ell)$ is the maximum over $P$, we choose the one $(k, \ell)$ such that $B_{L \cup S}^{k,\ell}$ has the fewest number of admissible edges. This is motivated by solution varieties of maximum matching problem; Given a bipartite graph, maximum matching is not necessarily unique. It is natural to expect that, the less admissible edges, the less maximum matchings; in our case, it may yield less maximum independent sets. Choosing $(k, \ell)$ in that way, we aim to fix the independent set of the subgraph that must have the fewest solution varieties.

For (ii), we pay attention to a node $(v_1, v_2, v_3)$ in $G_{L \cup S}^{k,\ell}$ that corresponds to a mandatory edge $(v_\kappa, v_{\kappa'})$ in $B_{L \cup S}^{k,\ell}$. We call such a node mandatory (in $G_{L \cup S}^{k,\ell}$) since it belongs to any maximum independent set there. As illustrated in [5], mandatory nodes should be treated with a greater care to approach a better solution. Specifically, when a maximum independent set of a certain subgraph is appended to the solution, some nodes and edges are removed from the entire graph. If a mandatory node in other subgraph is removed in this process, then the matching number of that subgraph decreases, which may also decrease the quality of the final solution.

---

**Algorithm 2** MATCHING for the PLSE problem

---
1: $S \leftarrow \emptyset, P = [3] \times [n]$
2: **while** $P$ is not empty **do**
3:     **repeat**
4:         $N \leftarrow$ the set of all the isolated nodes in $G_{LUS}^{ME}$
5:         $S \leftarrow S \cup N$
6:     **until** $N$ is empty
7:     choose the pair $(k, \ell)$ such that $f_{LUS}(k, \ell) = \max_{(k', \ell') \in P} f_{LUS}(k', \ell')$
              ▷ If there are plural candidates, choose $(k, \ell)$ such that $|AE(B_{LUS}^{k,\ell})|$ is the minimum.
8:     $I \leftarrow$ a maximum independent set in $G_{LUS}^{k,\ell}$
9:     $S \leftarrow S \cup I, P \leftarrow P \setminus \{(k, \ell)\}$
10: **end while**

---

Our idea is to append somewhat mandatory nodes to $S$ before a maximum independent set of a certain subgraph. Let $V_{LUS}^{ME} \subseteq V_{LUS}$ denote the set of all the mandatory nodes in $G_{LUS}^{k,\ell}$'-s over $(k, \ell)$'-s in $P$. Also let $G_{LUS}^{ME}$ denote the subgraph of $G_{LUS}$ that is induced by $V_{LUS}^{ME}$. Any independent set $N \subseteq V_{LUS}^{ME}$ in $G_{LUS}^{ME}$ can be appended to $S$, but in contrast to the subgraph $G_{LUS}^{k,\ell}$ on a facet, this $G_{LUS}^{ME}$ is not necessarily bipartite. Thus instead of a maximum independent set in $G_{LUS}^{ME}$, we use the set of all the isolated nodes in $G_{LUS}^{ME}$ as the set $N$ of nodes to be appended. We repeat updating $S \leftarrow S \cup N$ until $N$ is empty. In [5], Haraguchi et al. proposed to use a maximal (not necessarily maximum) independent set as $N$, which is generated by a minimum-degree greedy algorithm for the MIS problem. In our preliminary experiments, however, this new setting yields slightly better results.

Finally we summarize the algorithm MATCHING in Algorithm 2. The algorithm runs in $O(n^{11/2})$ time. The computation time is dominated by the iteration between Lines 3 and 6. The iteration times is $O(n^2)$ since the solution size is at most $n^2$. In each iteration, the most time-consuming task is to compute maximum matching for all the bipartite graphs $B_{LUS}^{k,\ell}$'-s to construct the graph $G_{LUS}^{ME}$. There are at most $3n$ bipartite graphs. Then we have $O(n^2) \times O(n) \times \tau_n = O(n^{11/2})$.

## 4 Hardest PLSE Instances

In this section, we show what are the hardest instances of the PLSE problem. Here the term "hard" is used to represent "time-consuming even by CPLEX." We evaluate the hardness of an instance according to whether or not CPLEX requires much computation time to find a global optimum. Any instance is parametrized by the grid length $n$ and the ratio $r \in [0, 1]$ of pre-assigned symbols over the $n \times n$ grid. We let $(n, r)$-*instance* be the general term that represents an $n \times n$ PLSE instance with a pre-assigned ratio $r$. Clearly, the larger $n$ is, the harder the instances tend to be. Then we are interested in how $r$ affects the hardness for a fixed $n$. We assert that, when $n$ is large to some extent, $(n, r)$-instance for $r \in [0.6, 0.7]$ should be the hardest among all. We draw this assertion based on computational studies.

CPLEX has optimization engines for both IP and CP. We denote by CPX-IP and CPX-CP the CPLEX engines for IP and CP, respectively. We evaluate the instance hardness by either engine as follows. Given $(n, r)$, we generate 100 instances by either of the two methods: *quasigroup completion (QC)* or *quasigroup with holes (QWH)*. We will explain these two methods soon. For each of the 100 instances, we run a CPLEX engine (i.e., CPX-IP or CPX-CP) to search for its global optimal solution. We set the time limit parameter to $6.0 \times 10^2$ seconds; if the CPLEX engine cannot find the global optimal solution within this time limit, we terminate the computation. Finally we count how many instances the CPLEX engine cannot find the global optimum within the time limit. We consider that, the more this number is, the harder the $(n, r)$-instance is for the CPLEX engine.

We explain the two methods for generating problem instances, QC and QWH. The methods are well-known in the field of constraint programming [3, 12]. QC starts with the empty assignment and assigns a symbol to an empty cell randomly so that the resulting assignment is a PLS. This is repeated until $\lfloor rn^2 \rfloor$ cells are assigned symbols. On the other hand, starting with an arbitrary $n \times n$ LS, QWH repeats removing the symbol from a randomly chosen cell until $\lfloor rn^2 \rfloor$ cells remain. Any QWH instance admits an LS as the optimal solution, whereas a QC instance does not necessarily so.

All the experiments are conducted by our PC that carries 2.80 GHz CPU and 4GB main memory. The version of CPLEX is 12.4. For CPX-CP, we set the level of default inference (`DefaultInferenceLevel`) and the level of all-different inference (`AllDiffInferenceLevel`) to `extended`, i.e., the most sophisticated constraint propagation technique is used. We set all the other parameters to default values.

We show the evaluation results in Fig. 1. As shown, whether the instance is generated by QWH or QC, whether the hardness is evaluated by CPX-IP or CPX-CP, and whether the grid size $n$ is small or large, the interval [0.6, 0.7] of $r$ is contained in the peak, or we dare to say the plateau, of a hardness "mountain." This supports our assertion that $(n, r)$-instance with $r \in [0.6, 0.7]$ should be the hardest.

Hereafter we treat only QC instances rather than QWH ones because the former ones are harder than the latter ones; under the fixed CPLEX engine and grid size $n$, the mountains in QC are larger than those in QWH in general.

Let us observe the two points of the hardest instances, i.e., $(n, r)$-QC instances with $r \in [0.6, 0.7]$: (i) whether the optimal solution is an LS or not, and (ii) whether there are a variety of the optimal solutions or not. For (i), when $r \geq 0.6$, we should not expect the optimal solution to be an LS. Whether an $(n, r)$-QC instance has an LS solution or not depends on $r$; when $r$ is small, the instance is "under-constrained" and thus must have an LS optimal solution, and when $r$ is large, the instance is "over-constrained" and hardly has an LS optimal solution. Our preliminary experiments for $10 \leq n \leq 20$ show that, when $r \geq 0.6$, none of 1000 random instances has an LS optimal solution, whereas more than 97 % of the instances have an LS

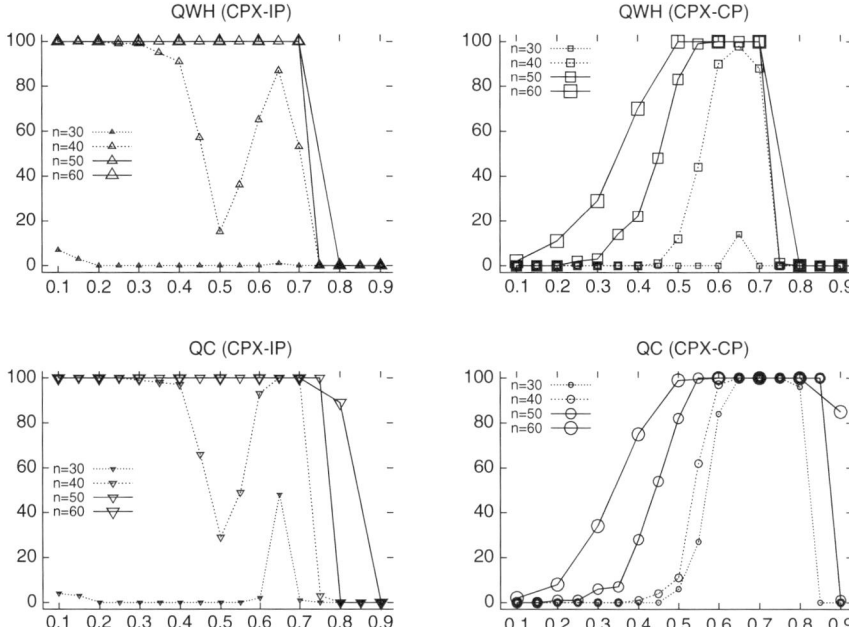

**Fig. 1** The number of instances (*vertical axis*) such that the CPLEX engines cannot find a global optimum within the time limit, with respect to the pre-assigned ratio (*horizontal axis*): The *upper/lower* two figures are for QWH/QC instances. The *left/right* two figures are for CPX-IP/CPX-CP

optimal solution when $r \leq 0.3$.[1] It is strongly expected that this tendency also holds for larger $n$.

For (ii), we presume that there are a variety of the optimal solutions in a hardest instance. Remark the two mountains that are observed when the hardness is evaluated by CPX-IP and $n = 40$. The instances on the left mountain ($r \leq 0.4$) are under-constrained instances that must have a variety of LS optimal solutions. The variety of LS optimal solutions should decrease when $r$ gets larger, and we drop into the valley between the two mountains (e.g., $r = 0.5$). When $r = 0.5$, CPX-IP fails to find a global optimum in 29 instances, which means that it is found in the rest 71 instances. We confirm that all of them are LS optimal solutions. This indicates that CPX-IP performs better on the instances that have a fewer variety of optimal solutions. When $r \geq 0.6$, the optimal solution is no longer an LS. Even so, inferring from the behavior of CPX-IP, we expect that the solution variety changes similarly as it changes from $r = 0$ to 0.6; the instances on the right mountain should have a variety of non-LS optimal solutions

---

[1] Gomes and Shmoys [3] claims that the boundary on whether the instance has an LS optimal solution or not should lie around $r = 0.42$.

The study in this section may give novel insight. The easy-hard-easy phase transition is a well-known phenomenon in the literature [3, 13–16]. The phenomenon clearly appears in the result of CPX-CP in Fig. 1. Gomes and Shmoys [3] observed the phase transition of the decision problem version of the PLSE problem, which asks to identify whether a given PLS has an LS extension or not; in our context, whether it has an LS optimal solution or not. They evaluate the instance hardness by the size of the search tree that a conventional backtracking algorithm induces. When the instance is generated by QC, the pre-assigned ratio $r$ from 0.4 to 0.5 yields the hardest ones, where there must be a boundary between extensible and non-extensible. When the instance is generated by QWH, it is always extensible and $r$ from 0.6 to 0.7 yields the hardest ones. Here we have evaluated the instance hardness in the context of optimization. We have observed that, whether the instance is generated by QC or QWH, it is more likely to be the hardest when $r \in [0.6, 0.7]$, for both IP and CP engines.

As is observed especially when $n \leq 40$, it is interesting to see that specialty is different between CPX-IP and CPX-CP; the former performs better (resp., worse) than the latter when $r \geq 0.6$ (resp., $r \leq 0.5$), where the instances are over-constrained (resp., under-constrained). Many IP/CP hybrid approaches are based on such characteristics of the two methodologies (e.g., [17, 18]).

## 5 Performance Evaluation

In this section, we illustrate how the algorithm MATCHING is effective, especially for the "hardest" instances that we addressed in the last section.

### 5.1 Comparison with CPLEX

We compare MATCHING with the two CPLEX engines, CPX-IP and CPX-CP. We show the ratio of the solution size that MATCHING outputs to the one that CPX-IP or CPX-CP outputs in Fig. 2. The average is taken over 100 instances. Similarly to the experiments so far, we set the time limit of the CPLEX engines to $6.0 \times 10^2$ seconds. When the computation time exceeds the time limit, we employ the best solution among those searched.

Let us focus on the "hardest" instances (i.e., $r = 0.6$ and 0.7). For these instances, CPLEX can hardly find the global optimum within the time limit (cf. Fig. 1). We see that MATCHING is likely to deliver a better solution than CPLEX especially when $n \geq 50$, although the computation time is much shorter.

We show the CPU time that MATCHING takes to deliver a solution in Fig. 3. We confirm that, in all the tested instances, only 5 seconds are needed for MATCHING to deliver a solution. Here we see the effectiveness of the heuristic algorithm; although

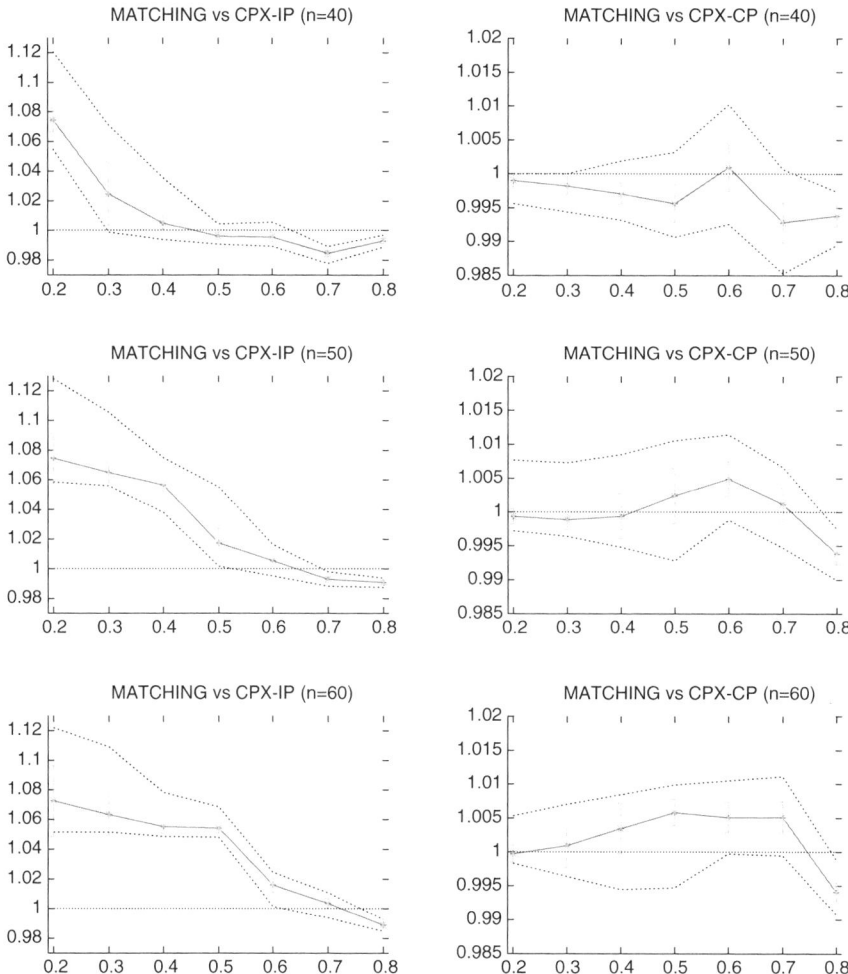

**Fig. 2** Performance comparison between MATCHING and CPLEX ($n = 40$, 50 and 60): In each chart, the *vertical axis* indicates the averaged ratio of the solution size of MATCHING over that of CPX-IP (*left*) or CPX-CP (*right*). An error-bar represents the standard deviation. Two *dotted curves* indicate the maximum and minimum ratios among the tested 100 instances. The *horizontal axis* indicates the pre-assigned ratio $r$

the computation time is less than 5 seconds, the MATCHING solution is better than the ones that the CPLEX engines output after $6.0 \times 10^2$ seconds.

As can be seen in Fig. 3, when $n$ is fixed, the smaller the pre-assigned ratio $r$ is, the more computation time MATCHING takes. This is because the graph $G_L$ in the MIS notation has more nodes and edges when $r$ is smaller. We show the averaged numbers of nodes and edges in Table 1. It is interesting to see that the instance hardness does

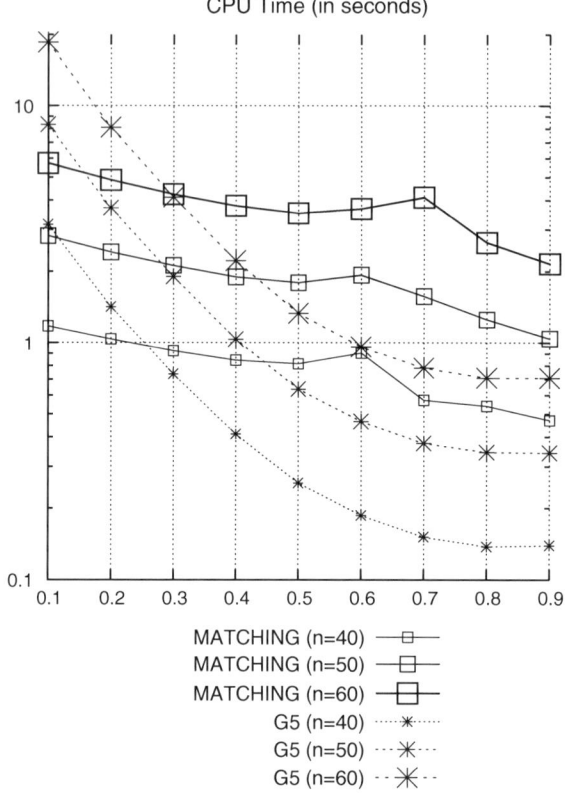

**Fig. 3** CPU time (in seconds, *vertical logarithmic axis*) of MATCHING and G5 with respect to the pre-assigned ratio $r$ (*horizontal axis*)

not necessarily depend on the graph size but may depend on the graph structure. In fact, some benchmark MIS instances that are recognized as "hard" in the literature (e.g., DIMACS challenge) consist of thousands of nodes or even fewer nodes.

## 5.2 Comparison with Heuristic Algorithms for MIS

CPLEX is based on exact algorithms. Thus in principle, its purpose is different from that of MATCHING, a heuristic algorithm, which is designed not for finding a global optimal solution but for generating a good solution quickly. Here we compare MATCHING with other heuristics that are developed for the MIS problem.

The paper [8] contains a collection of 11 heuristic algorithms for the MIS problem, all of which are under the concept of surrogate constraint based heuristics [7]. The 11 algorithms are named from G1 to G11 respectively, according to the sophistication

**Table 1** The size of the graph $G_L = (V_L, E_L)$ in the MIS formulation: The density refers to the ratio $|E_L|/\binom{|V_L|}{2}$ in percent

|         |           | $r = 0.2$ | 0.6 | 0.7 | 0.8 |
|---------|-----------|-----------|-----|-----|-----|
| $n = 40$ | $|V_L|$   | $3.3 \times 10^4$ | $4.4 \times 10^3$ | $2.0 \times 10^3$ | $6.3 \times 10^2$ |
|         | $|E_L|$   | $1.2 \times 10^6$ | $4.4 \times 10^4$ | $1.1 \times 10^4$ | $1.7 \times 10^3$ |
|         | (density) | 0.23 % | 0.44 % | 0.60 % | 0.86 % |
| 50      | $|V_L|$   | $6.5 \times 10^4$ | $8.5 \times 10^3$ | $3.8 \times 10^3$ | $1.2 \times 10^3$ |
|         | $|E_L|$   | $3.1 \times 10^6$ | $1.1 \times 10^5$ | $2.7 \times 10^4$ | $4.0 \times 10^3$ |
|         | (density) | 0.15 % | 0.29 % | 0.38 % | 0.57 % |
| 60      | $|V_L|$   | $1.1 \times 10^5$ | $1.5 \times 10^4$ | $6.4 \times 10^3$ | $2.0 \times 10^3$ |
|         | $|E_L|$   | $6.4 \times 10^6$ | $2.2 \times 10^5$ | $5.5 \times 10^4$ | $8.0 \times 10^3$ |
|         | (density) | 0.10 % | 0.20 % | 0.27 % | 0.40 % |

level of the idea, i.e., the larger the index is, the more sophisticated the idea is. Among these, we use G1 and G5 as our competitors.

G1  A typical minimum-degree greedy algorithm. Starting from an empty solution $S = \emptyset$, it repeats choosing the node $v$ that has the minimum degree $|N(v)|$ and appending $v$ to $S$. Once $v$ is appended to $S$, $v$ itself, all the neighbors of $v$ (i.e., the nodes in $N(v)$), and all the incident edges to these nodes are removed from the graph. The repetition ends when no node remains in the graph.

G5  A "look-ahead" minimum-degree greedy algorithm. It is an extension of G1; when there are more than one node that have the minimum degree, it chooses such $v$ that maximizes $|\bigcup_{u \in N(v)} N(u)|$, i.e., the one such that most edges are removed from the graph.

It is reported that G5 and G11 are the best among the 11 algorithms. G5 is a simple algorithm as above, whereas G11 requires parameter search to enhance its performance. We use G1 just for a benchmark.

We show the comparison results in Fig 4. When $r = 0.8$, MATCHING is worse than both G1 and G5 in all cases. However, MATCHING outperforms G1 for any $r \leq 0.7$. Let us concentrate on the comparison between MATCHING and G5. We see that they are quite competitive for non-hardest instances (i.e., $r \leq 0.5$). For hardest instances (i.e., $r = 0.6$ and 0.7), when $n = 40$, MATCHING is competitive with ($r = 0.6$) or even inferior ($r = 0.7$) to G5, but when $n \geq 50$, it wins over G5. Taking into account that MATCHING performs relatively better than CPLEX when $n$ is larger, it is expected to be effective for hardest instances especially when $n$ is large.

Figure 3 compares the CPU time between MATCHING and G5. The computation time of G5 depends on the graph size more than MATCHING. When $r \leq 0.2$, G5 runs slower than MATCHING, but when $r > 0.2$, G5 becomes faster than MATCHING. When $0.3 \leq r \leq 0.4$, one may apply G5 rather than MATCHING since they are competitive in performance but the former runs faster than the latter. When $r \geq 0.8$, it is remarkable that G5 outperforms MATCHING both in performance and in computation time. G5

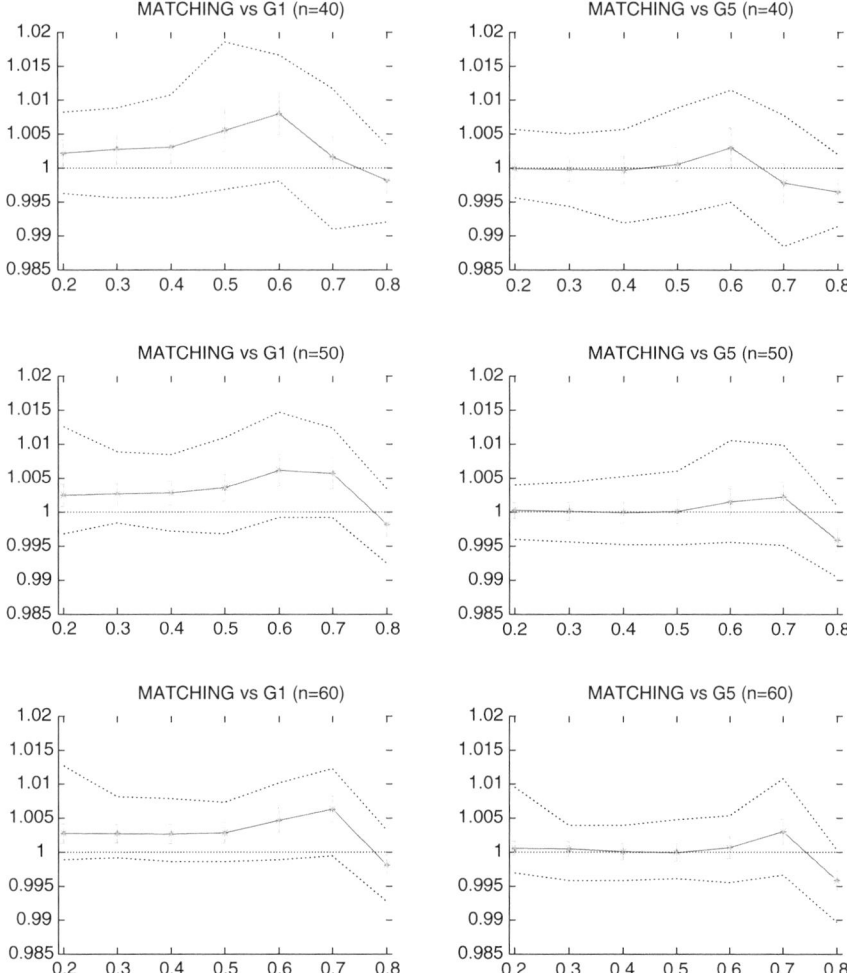

**Fig. 4** Performance comparison between MATCHING and surrogate constraint based heuristics ($n = 40$, $50$ and $60$): In each chart, the *vertical axis* indicates the averaged ratio of the solution size of MATCHING over that of G1 (left) or G5 (*right*). An error-bar represents the standard deviation. Two *dotted* curves indicate the maximum and minimum ratios among the tested 1000 instances. The *horizontal axis* indicates the pre-assigned ratio $r$

is recognized as one of the best algorithms in [8] for general MIS problem instances. We have observed that it also performs relatively well on the PLS instances in MIS formulation.

# 6 Concluding Remarks

In this paper, we have addressed the heuristic algorithm MATCHING for the PLSE problem that was proposed in [5]. We have presented how it is effective for the hardest instances in comparison with CPLEX and other heuristics.

Below we enumerate future work. We should investigate the reason why MATCHING is more effective than others for the hardest instances. We have not embedded any explicit mechanism that lets the algorithm be effective for the hardest instances; it happens to be effective. The reason should be clarified for further improvement.

We have obtained an effective constructive algorithm for the PLSE problem. Then it is natural to consider applying *metaheuristics* as the next step. Local search is one of the most fundamental technique in metaheuristics, and thus should be applied at first. We have already worked on this, utilizing the recent result of Andrade et el. [19] on how to implement local search efficiently for the MIS problem. The preliminary experiments show that, given the same initial solution and the same time limit, our local search algorithm delivers a better solution than CPLEX and LocalSolver [20]. We will present these results in our future papers.

The theoretical aspects are also interesting. Hajirasouliha et al. [21] pointed out that the PLSE problem is a special case of 3-set packing problem, and showed local search to provide a $(2/3 - \varepsilon)$-approximation algorithm for any $\varepsilon > 0$.[2] They used Hurkens and Schrijver's well-known result on $k$-set packing problem [22] such that local search provides a $(2/k - \varepsilon)$-approximation algorithm. Recently, Cygan [23] improved the approximation ratio for the $k$-set packing problem from $2/k - \varepsilon$ to $3/(k + 1) - \varepsilon$ by means of *bounded pathwidth local search*. Accordingly, the best known approximation ratio for the PLSE problem now becomes $3/4 - \varepsilon$. On the other hand, the PLSE problem is an APX-hard problem [21]. It is interesting future work to improve the approximation ratio to pursuit the limit.

We are not aware of the application areas of such a large PLSE instance as treated in this paper. Nevertheless, it is significant to accumulate the research results in the literature as we may encounter large and "hard" instances in the nearest future. For such instances, our algorithm may help generate a good solution quickly rather than state-of-the-art optimization solvers. Besides, since PLSE is a special case of MIS or $k$-set packing problems, once we have obtained certain research results on PLSE, we may extend them to MIS or set packing, and vice versa. This is a reasonable approach to navigate the vast and deep ocean of NP-hard problems.

---

[2] The larger the extent of the neighborhood is, the smaller $\varepsilon$ becomes.

# References

1. Colbourn, C.J.: The complexity of completing partial Latin squares. Discrete Appl. Math. **8**, 25–30 (1984)
2. Barry, R.A., Humblet, P.A.: Latin routers, design and implementation. IEEE/OSA J. Lightwave Technol. **11**(5), 891–899 (1993)
3. Gomes, C.P., Shmoys, D.: Completing quasigroups or Latin squares: a structured graph coloring problem. In: Proc. Computational Symposium on Graph Coloring and Generalizations (2002)
4. Kumar, R., Russel, A., Sundaram, R.: Approximating latin square extensions. Algorithmica **24**(2), 128–138 (1999)
5. Haraguchi, K., Ishigaki, M., Maruoka, A.: A maximum matching based heuristic algorithm for partial latin square extension problem. In: Proc. FedCSIS **2013**, 347–354 (2013)
6. IBM, ILOG CPLEX: http://www-01.ibm.com/software/commerce/optimization/cplex-optimizer/
7. Glover, F.: Tutorial on surrogate constraint approaches for optimization in graphs. J. Heuristics **9**, 175–227 (2003)
8. Alidaee, B., Kochenberger, G., Wang, H.: Simple and fast surrogate constraint heuristics for the maximum independent set problem. J. Heuristics **14**, 571–585 (2008)
9. Hopcroft, J.E., Karp, R.M.: An $n^{5/2}$ algorithm for maximum matchings in bipartite graphs. SIAM J. Comput. **2**(4), 225–231 (1973)
10. Cymer, R.: Dulmage-Mendelsohn canonical decomposition as a generic pruning technique. Constraints **17**, 234–272 (2012)
11. Garey, M.R., Johnson, D.S.: Computers and intractability: a guide to the theory of NP-completeness. W. H. Freeman & Company, New York (1979)
12. Barták, R.: On generators of random quasigroup problems. In: Proc. CSCLP **2005**, 164–178 (2006)
13. Gomes, G., Selman, B.: Problem structure in the presence of perturbations. In: Proc. AAAI 97, pp. 221–227 (1997)
14. Shaw, P., Stergiou, K., Walsh, T.: Arc consistency and quasigroup completion. In: Proc. ECAI-98 (workshop on binary constraints) (1998)
15. Monasson, R., Zecchina, R., Kirkpatrick, S., Selman, B., Troyansky, L.: Determining computational complexity from characteristic phase transitions. Nature **400**, 133–137 (1999)
16. Ansotegui, C., Bejar, R., Fernandez, C., Mateu, C.: On the hardness of solving edge matching puzzles as sat or csp problems. Constraints **18**, 7–37 (2013)
17. Appa, G., Magos, D., Mourtos, I.: Searching for mutually orthogonal Latin squares via integer and constraint programming. Eur. J. Oper. Res. **173**(2), 519–530 (2006)
18. Fontaine, D., Michel, L.: A high level language for solver independent model manipulation and generation of hybrid solvers. In: Proc. CPAIOR 2012, pp. 180–194 (2012)
19. Andrade, D., Resende, M., Werneck, R.: Fast local search for the maximum independent set problem. J. Heuristics **18**, 525–547 (2012)
20. LocalSolver: http://www.localsolver.com/
21. Hajirasouliha, I., Jowhari, H., Kumar, R., Sundaram, R.: On completing Latin squares. In: Proc. STACS 2007, pp. 524–535
22. Hurkens, C.A.J., Schrijver, A.: On the size of systems of sets every t of which have an SDR, with an application to the worst-case ratio of heuristics for packing problems. SIAM J. Discrete Math. **2**(1), 68–72 (1989)
23. Cygan, M.: Improved approximation for 3-dimensional matching via bounded pathwidth local search. arXiv preprint arXiv:1304.1424 (2013)

# Branch and Price for Preemptive and Non Preemptive RCPSP Based on Interval Orders on Precedence Graphs

**Aziz Moukrim, Alain Quilliot and Hélène Toussaint**

**Abstract** This paper first describes an efficient exact algorithm to solve Preemptive RCPSP and next discusses its extension to Non Preemptive RCPSP. In case of Preemptive RCPSP, we propose a very original and efficient branch and bound/price procedure based upon minimal interval order enumeration, which is implemented with the help of the generic SCIP software. We perform tests on the famous PSPLIB instances which provide very satisfactory results. To the best of our knowledge it is the first algorithm able to solve at optimality all the set of j30 instances of PSPLIB in a preemptive way. The two last sections are devoted to the description of some heuristics, which also involve the interval order framework and the basic antichain linear program and which aim at handling larger scale RCPSP preemptive instances, and to a discussion of the way our algorithm may be extended to Non Preemptive RCPSP.

**Keywords** Preemptive RCPSP · Branch and price

## 1 Introduction

This paper mainly deals with the *Preemptive Resource Constrained Project Scheduling Problem* (RCPSP: see [1, 2]). RCPSP aims at scheduling a set of activities, submitted to precedence and resource constraints, while minimizing the induced

A. Moukrim
Université de Technologie de Compiègne, Heudiasyc, CNRS UMR 7253,
60203 Compiègne, France
e-mail: aziz.moukrim@hds.utc.fr

A. Quilliot (✉) · H. Toussaint
LIMOS CNRS UMR 6158 LABEX IMOBS3, Université Blaise Pascal,
63173 Aubière, France
e-mail: quilliot@isima.fr

H. Toussaint
e-mail: toussain@isima.fr

© Springer International Publishing Switzerland 2015
S. Fidanova (ed.), *Recent Advances in Computational Optimization*,
Studies in Computational Intelligence 580, DOI 10.1007/978-3-319-12631-9_6

*makespan* (total duration of the project) value. The precedence constraints mean that some activities must be completed before others can start. The resource constraints specify that each activity requires constant amounts of renewable resources during all the time it is processed, these resources having limited capacities. This problem has been extensively studied in its non preemptive version [3–6], which means that every activity has to be run as a whole, without any kind of interruption. There exist several variants of RCPSP (see [7, 8] for recent surveys, some of them involving uncertainty [9]). We talk about *Preemptive* RCPSP when an activity may be run in several steps: one may launch such an activity, interrupt it, keep on with this activity a little further, and so on. There exists few works on Preemptive RCPSP: [10] developed a branch and bound algorithm, [11] proposed a tree search procedure augmented with pruning rules (best-first tree search), [12] proposed an integer linear program which add preemption penalties, [13, 14] dealt with preemption in an heuristic way and [15] and also [16] designed a genetic algorithm for multi-mode *Preemptive* RCPSP.

The approach which we propose here is a Branch and Bound one which involves constraint propagation, as well as the management of specific rational *Antichain* linear program whose variables are associated with subsets of activities which may be simultaneously processed during the schedule. This LP, which was first introduced by [17], provides us with a lower bound of both *Preemptive* and *Non Preemptive* RCPSP, which was also used in a successful way in [6, 18]. But dealing with it requires implementing a pricing or column generation scheme. It was proved in [19] that if the input RCPSP instance satisfies some ad hoc properties, then any optimal solution of the *Antichain* linear program may be turned into a feasible optimal schedule, without any increase of the *makespan* value. What we do here is to use this property in order to perform a tree search which may be viewed as being embedded into the enumeration process of all minimal extensions of the precedence relation which define *interval* orders. The resulting process happens to be very efficient, since it solves in an exact way all 30 activity instances of the PSPLIB library, and to improve best existing lower bounds for several 60/120 activity instances of this library. We also use the same framework in order to derive fact heuristics and a proposal for the handling of the non preemptive case.

So the paper is organized as follows: we first recall what is *Preemptive* RCPSP (Sect. 2), and next introduce our basic theoretical tools related to the *Antichain* LP and to interval orders (Sect. 3). Section 4 describes the algorithm INT-ORD-ENUM and its implementation, and Sect. 5 is devoted to a presentation of experimental results. In Sect. 6, we describe several heuristics, which all involve the Interval Order framework and the *Antichain* linear program, and, in last Sect. 7, we explain how our *Interval Order* framework may be adapted to the *Non Preemptive* version of this scheduling problem.

## 2 Preemptive RCPSP

An instance $I = (X, K, \ll)$ of the *Resource Constrained Project Scheduling Problem* is defined by:

- A set $X = 1, \ldots, n$ of $n$ *activities*: $\forall i \in X$, $d_i$ denotes the *duration* of activity $i$
- A set $K = 1, \ldots, m$ of $m$ *resources*: $\forall i \in X$, $\forall k \in K$, $r_{ik}$ denotes the requirement of activity $i$ for resource $k$; those resources are given back to the system once the activity is over
- $\forall (i, j) \in X^2$, $i \ll j$ means that $i$ *precedes* $j$: activity $j$ cannot start before $i$ is over (*Precedence* constraints)

In the case of *Non Preemptive* RCPSP, scheduling only means computing the starting times $t_i$, $i \in X$, of the activities. A schedule $\sigma = (t_i, i \in X)$ is feasible if it satisfies:

- the *Precedence* constraints
- the *Resource* constraints: at any time $t$ during the process, and for any resource $k$, the sum $\sum_{i \in Act(\sigma,t)} r_{ik}$ does not exceed the global resource amount $R_k$, $Act(\sigma, t) = \{i \text{ such that } t_i \leq t < t_i + d_i\}$ denoting the set of the activities currently run at time $t$ according to schedule $\sigma$.

So, solving *Non Preemptive* RCPSP means computing $\sigma$ with a minimal *makespan* (total duration of the process). In case preemption is allowed, scheduling an activity $i$ means first decomposing $i$ into a sequence of sub-activities $i_1, \ldots, i_{h(i)}$, with durations $d_{i_1}, \ldots, d_{i_{h(i)}}$, such that: $\sum_{q=1}^{h(i)} d_{i_q} = d_i$, and next scheduling all these sub-activities in the sense of standard RCPSP. Since there does not exist any "a priori" restriction either on the number of sub-activities or on their durations, which may be arbitrarily small, the existence of an optimal solution of Preemptive RCPSP has to be discussed.

## 3 Fundamental Tools

### 3.1 The Antichain Linear Program: A Lower Bound

Let $I = (X, K, \ll)$ be some Preemptive RCPSP instance, defined according to notations of Sect. 2. We suppose (we clearly may do it) that precedence relation $\ll$ is transitive. Then we define an *antichain* as being any subset $a$ of $X$ such that there does not exist $(i, j) \in a^2$ such that $i \ll j$. We say that such an antichain is *valid* if: $\forall k \in K$, $\sum_{i \in u} r_{ik} \leq R_k$. It comes that a subset $u \subseteq X$ of activities is a *valid antichain* iff activities in $a$ may be simultaneously run inside some feasible schedule. We denote by $\mathcal{A}$ the set of all *valid antichains*. Then we set the following *Antichain Linear Program* $(\mathcal{P})_{Ant}$ associated with instance $I = (X, K, \ll)$, which was already introduced in [17, 19]:

$$\text{Minimize} \sum_{a \in \mathcal{A}} z_a$$

s.t.

$$(\mathcal{P})_{Ant} \quad \forall i \in X, \sum_{a \in \mathcal{A} | i \in a} z_a = d_i \quad (C1)$$

$$\forall a \in \mathcal{A}, z_a \geq 0$$

**Explanation**: if $\sigma$ is any feasible schedule related to instance $I$, we may associate with $\sigma$ and with any valid antichain $a$, the total amount of time $z(\sigma)_a$ during which the activities which are simultaneously run according to $\sigma$ are exactly the activities of $a$. We see that $z(\sigma) = (z(\sigma)_a, a \in A)$ is a feasible solution of $(\mathcal{P})_{Ant}$ since constraints $(C1)$ say that any activity $i$ has to be completely done, or, equivalently, that the duration of all antichains containing $i$ must be equal to the duration of $i$. So, the optimal value of $(\mathcal{P})_{Ant}$ provides us with a lower bound of the optimal value of $I$, which we denote by $LB(I)$.

## 3.2 Dealing with $(\mathcal{P})_{Ant}$: Column Generation

Since the set $\mathcal{A}$ may be very large, even when the activity set $X$ is small, we need to handle the *Antichain* LP$(\mathcal{P})_{Ant}$ through *column generation*. *Column Generation* consists in initializing this LP with a few number of *active* variables (which may from application of some heuristic), and then in iteratively solving the induced *restricted problem* at optimality and using the dual variables to generate a new improving primal variable. The search for this improving primal variable is called the related *Pricing* Problem. When this technique is associated to a Branch and Bound process (usually for integer formulation) it gives rise to a *Branch and Price* process. In our case, let us consider some active antichain subset $B \subseteq \mathcal{A}$, together with some dual solution $\lambda$ of the restricted LP $(\mathcal{P})_{Ant}^B$ defined by $B$:

$$\text{Minimize} \sum_{a \in B} z_a$$

s.t.

$$(\mathcal{P})_{Ant}^B \quad \forall i \in X, \sum_{a \in B | i \in a} z_a = d_i \quad (C1)$$

$$\forall a \in B, z_a \geq 0$$

Then solving the related pricing problem $PRICE(\lambda)$ means computing some valid antichain $a$, such that: $\sum_{i \in a} \lambda_i > 1$. Though this problem is NP-Complete, it may be efficiently handled through a combination of greedy search and Integer Linear

Programming (LIP). A well-fitted LIP formulation $L$-$PRICE(\lambda)$ of $PRICE(\lambda)$ comes as follows:

$$\text{Maximize} \sum_{i \in X} \lambda_i y_i$$

s.t.

$$L\text{-}PRICE(\lambda) \quad \forall (i, j) \in X^2 | i \ll j, \quad y_i + y_j \leq 1 \quad\quad (C2)$$

$$\forall k \in K, \quad \sum_{i \in X} r_{ik} y_i \leq R_k \quad\quad (C3)$$

$$\forall i \in X, \quad y_i \in \{0, 1\}$$

## 3.3 Turning a Solution of $(\mathcal{P})_{Ant}$ into a Feasible Schedule?

Unfortunately, Linear Program $(\mathcal{P})_{Ant}$ only provides us with a lower bound of Preemptive RCPSP instance $I$: if vector $z = (z_a, a \in A)$ is a feasible solution of $(\mathcal{P})_{Ant}$, it may not be possible to turn it into a feasible solution of $I$. As a matter of fact, we may provide the valid antichain set $\mathcal{A}$ with an oriented graph structure $(\mathcal{A}, E_\ll)$ by setting that there exists an arc $(a, b) \in E_\ll$ from antichain $a$ to antichain $b$, if there exist activities $i \in a$ and $j \in b$, such that $i \ll j$. Then we easily check that:

**Theorem 1** *Let $z$ be some feasible solution of $(\mathcal{P})_{Ant}$, and $\mathcal{A}(z) \subseteq \mathcal{A}$ be the set $\mathcal{A}(z) = \{a \in \mathcal{A}$ such that $z_a \neq 0\}$ of active antichains according to $z$. Then there exists a feasible schedule $\sigma$ such that $z = z(\sigma)$ if and only if the subgraph $(\mathcal{A}(z), E_\ll)$ does not contain any circuit.*

*Proof* Left to the reader.

Still, we may notice that program $(\mathcal{P})_{Ant}$ provides us with additional understanding of Preemptive RCPSP: if $\sigma$ is any feasible Preemptive RCPSP schedule, if $z(\sigma) = (z(\sigma)_a, a \in A)$ is the related solution of $(\mathcal{P})_{Ant}$, and if $\mathcal{A}(z(\sigma))$ is the related active antichain set, then one sees that solving the restricted linear program $(\mathcal{P})_{Ant}^{\mathcal{A}(z(\sigma))}$ through Primal Simplex Algorithm provides us with another feasible schedule $\sigma^*$ with makespan no larger than the makespan of $\sigma$. Moreover, Linear Programming Theory tells us that the number of active antichains related to $\sigma^*$, that means the cardinality of $\mathcal{A}(z(\sigma^*))$ does not exceed the number of constraints of $(\mathcal{P})_{Ant}^{\mathcal{A}(z(\sigma))}$, which is equal to the cardinality of the activity set $X$. This makes appear Preemptive RCPSP as a combinatorial problem related to the search of some acyclic subgraph $(B, F_\ll)$ of the antichain graph $(\mathcal{A}, E_\ll)$, such that $Card(B) \leq Card(X)$ and that the optimal value of the program $(\mathcal{P})_{Ant}^B$ is minimal. This confirms the existence of an optimal solution.

Also, we may notice that no activity which is not in the set $Min(X)$ of the activities which are minimal in the sense of the precedence relation $\ll$, may start before the time when at least one activity in $Min(X)$ is completed. We deduce that the lower bound which derives from the $(\mathcal{P})_{Ant}$ program may be improved by adding the following constraint: $\sum_{a \in A_{Min}} z_a \geq Inf(d_i, i \in Min(X))$, with $A_{Min} = a \in \mathcal{A}$, such that $a \subseteq Min(X)$.

We denote by $LB^*(I)$ this improved lower bound.

## 3.4 Interval Orders

A partially ordered set $(Z, <)$ is an *interval order* if the elements $z$ of $Z$ may be represented as closed intervals $[o(z), d(z)]$ of the real line, in such way that, for any pair $z, z'$ in $Z$: $z < z'$ if and only if $d(z) < o(z')$. It is known (see [20]), that the partially ordered set $(Z, <)$ is an interval order if and only if there does not exists $x, y, z, t \in Z$ such that:

- $x < y$ and $t < z$; (C4)
- there does not exist any other pair $u < v$ with $u, v \in \{x, y, z, t\}$ than the pairs in (C4) above.

  Figure 1 shows the forbidden pattern associated with interval orders.
  If we consider now our Preemptive RCPSP instance $I = (X, K, \ll)$, we see that:

**Theorem 2** *If the partial order $(X, \ll)$ is an interval order, then the oriented antichain graph $(\mathcal{A}, E_\ll)$ is acyclic (does not contain any circuit).*

*Proof* We suppose the converse, and consider some circuit $\Gamma$ in $(\mathcal{A}, E_\ll)$ with minimal length. Then we must distinguish two cases:

- **first case**: Length$(\Gamma) = 2$, which means that $\Gamma$ contains two antichains $a$ and $b$. Then we see that there must exist $i_1, j_2 \in a$, $i_2, j_1 \in b$ such that: $i_1 \ll j_1$ and $i_2 \ll j_2$. Then it becomes easy to check that $i_1, j_1, i_2, j_2$ define a forbidden pattern in the above sense, which induces a contradiction.
- **second case**: Length$(\Gamma) \geq 3$, which means that contains 3 consecutive antichains $a, b, c$, and that there must exist $x \in a$, $y, z \in b$, $t \in c$, such that $x \ll y$ and $z \ll t$. But we also deduce from the minimality of Length$(\Gamma)$ and from the fact that $a, b, c$ are antichains that $x, y, z, t$ must define a forbidden pattern in the above sense, which induces again a contradiction.

  This result will impact the rest of the paper. Clearly, if $\sigma$ is a feasible schedule for the Preemptive RCPSP instance $I = (X, K, \ll)$, it is possible to extend the precedence relation $\ll$ into an interval order $\ll_\sigma$, in such a way $\sigma$ remains consistent

**Fig. 1** Interval order forbidden pattern

with $\ll_\sigma$. So, Theorem 2 makes appear that we only need, in order to deal instance $I$, to enumerate the extensions $\ll^*$ of $\ll$ which are interval orders. As a matter of fact, we may restrict ourselves to those extensions $\ll^*$ which are minimal for inclusion.

# 4 The Branch/Bound Algorithm INT-ORD-ENUM

## 4.1 A Reformulation of Preemptive RCPSP Instance I

Sections 2 and 3 lead us to reformulate *Preemptive RCPSP* instance $I = (X, K, \ll)$:

**Preemptive RCPSP Reformulation:** {Compute an extension $\ll^*$ of the precedence relation $\ll$ which is an interval order and which is such that, if $z^*$ is an optimal solution of the related LP $(\mathcal{P})_{Ant}$, obtained through Primal Simplex Algorithm and column generation, the optimal value $1.z^*$ of is the smallest possible.}

So, our algorithm INT-ORD-ENUM is a Branch/Bound algorithm, which performs some enumeration of the extensions $\ll^*$ of $\ll$. We must now specify the main components of such a tree search process, which are about:

- the extensions of Preemptive RCPSP instance $I = (X, K, \ll)$ which define the nodes of the related search tree;
- the way branching is performed;
- the way bounding and related filtering are performed;
- the way constraint propagation is performed;
- the branching strategy;
- the way the whole algorithm is implemented.

## 4.2 The Nodes of the INT-ORD-ENUM Search Tree

A node of the search tree induced by a branch/bound algorithm is usually defined by a set of additional constraints imposed to the initial problem. In the case of the Preemptive RCPSP instance $I = (X, K, \ll)$, those constraints are:

- additional precedence constraints $i \ll j$;
- anti-precedence constraints $i \dashrightarrow j: i \dashrightarrow j$ means that $i \ll j$ is forbidden.

So, we may identify any node of the search tree with a pair $(Add_\ll, Add_{\dashrightarrow})$, where $Add_\ll$ and $Add_{\dashrightarrow}$ are respectively the sets of additional precedence constraints and anti-precedence constraints which constrain $\ll^*$ as follows:

- $(\ll \cup Add_\ll) \subseteq \ll^*$;
- $(Add_{\dashrightarrow} \cap \ll^*) = Nil$.

Clearly, if current $\ll^*$ defines an interval order, the related node is a terminal node. Following Fig. 2 illustrates this branching process.

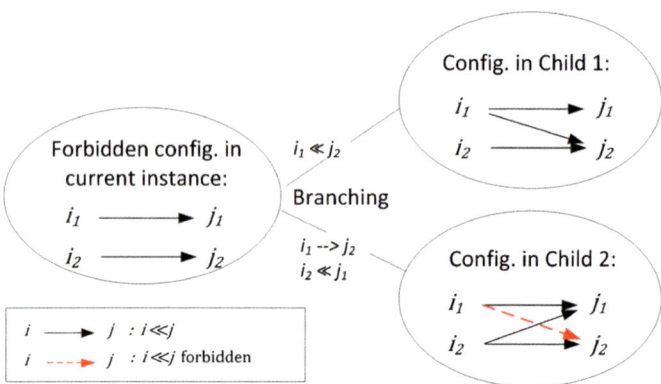

**Fig. 2** The branching mechanism

## 4.3 The Branching Mechanism

If current precedence relation, managed in such a way it always remains transitive, is not an interval order, then it must contain some forbidden pattern $i_1, j_1, i_2, j_2$:

- $i_1 \ll j_1$ and $i_2 \ll j_2$; (C5)
- no other pair $u \ll v$ exists with $u, v \in \{i_1, i_2, j_2, j_1\}$ than the pairs in (C5) above.

This forbidden pattern allows us to perform a binary branching process by successively considering the 2 following alternatives:

- 1 th alternative (1 th son): insert $i_1 \ll j_2$ into $Add_{\ll}$;
- 2 th alternative (2 th son): insert $i_2 \ll j_1$ into $Add_{\ll}$ and insert $i_1 \dashrightarrow j_2$ into $Add_{\dashrightarrow}$.

## 4.4 Lower Bound, Upper Bound and Related Filtering

*Lower Bound* The lower bound which derives from a current node defined by a pair $(Add_{\ll}, Add_{\dashrightarrow})$, is provided by the optimal value of the program $(\mathcal{P})_{Ant}$, where valid antichains are considered as deriving from $(\ll \cup Add_{\ll})$. This problem is handled through column generation, as explained in Sect. 3.3, and the Pricing problem $PRICE(\lambda)$ is handled while using the ILP model of Sect. 3.3.

*Upper Bound* Also, we make in such a way that we are provided, as part of a pretreatment, with an initial upper bound *UB:* in order to get this initial upper bound, we apply to instance $I$, a greedy randomized algorithm designed for the Non Preemptive RCPSP (see [20]) and which, in case of 30 activity PSPLIB instances, approximates the optimal *Non Preemptive* RCPSP optimal value by less than 2 % in average. *UB* is updated as soon as some feasible solution is computed.

*Related Filtering* Of course, if the solution $z^*$ of LP $(\mathcal{P})_{Ant}$, is such that the subgraph $(\mathcal{A}(z^*), E_\ll)$ does not admit any circuit, we consider that we have been reaching some terminal node of the search tree. In case related value $1.z^*$ is smaller than the value of the current solution (current upper bound $UB$ of the forthcoming Sect. 4.5, we update this current solution as a feasible schedule $\sigma$ such that $z^*=z^*(\sigma)$.

## 4.5 Constraint Propagation

We apply several kind of inference rules $\alpha \vDash \beta$, whose semantics come as follows: $\alpha$ is the precondition part, and $\beta$ is the consequent part (additional relations inserted into sets $Add_\ll$ and $Add_{\dashrightarrow}$.). The first class of rules deals with transitivity, and makes in such a way that, at any time, current relation $\ll^* = (\ll \cup Add_\ll)$ remains transitive:

Rule 1    $i \ll^* j, z \ll^* i \vDash z \ll^* j$;
Rule 1′    $i \ll^* j, j \ll^* z \vDash i \ll^* z$;

Of course, any relation $i \ll i$ induces a *Failure* signal.

     The second one deals in a classical way with largest paths and current upper bound $UB$. We add two dummy activities: $s$ (source) and $p$ (sink) defined as usual and, at every time during the process, we are provided, for every activity $i$, with: $\pi(i) =$ earliest finish time for $i$ ; $\Pi(i) =$ latest starting time for $i$. The following classical inference rules keep the current precedence relation $(\ll \cup Add_\ll)$ from inducing the existence of a largest path with length $\geq UB$:

Rule 2    $\pi(i) = \alpha, i \ll^* y$ and $\alpha + d_y > \pi(y) \vDash \pi(y) = \alpha + d_y$;
Rule 2′    $\Pi(i) = \alpha, y \ll^* i$ and $\alpha + d_y > \Pi(y) \vDash \Pi(y) = \alpha + d_y$;
Rule 3    $\pi(i) = \alpha, \alpha + \Pi(y) > UB \vDash i \dashrightarrow y$;
Rule 3′    $\Pi(i) = \alpha, \alpha + \pi(y) > UB \vDash y \dashrightarrow i$;

Rules 3 and 3′ forbid any additional precedence relation which would induce the existence of a largest path with length $\geq$ UB to be inserted into $Add_\ll$.

Rule 4    $\pi(i) = \alpha, UB - \Pi(y) + d_y \leq \alpha - d_i \vDash y \ll^* i$;
Rule 4′    $\Pi(i) = \alpha, UB - \alpha + d_i \leq \pi(y) - d_y \vDash i \ll^* y$;

Rules 4 and 4′ insert into $Add_\ll$. Additional precedence relations which should be satisfied in any schedule with makespan no more than $UB$.

     The last class of rules deals with the forbidden patterns of Sect. 3.4, and aims at keeping current relation $(\ll \cup Add_\ll)$ from containing any such a pattern:

Rule 5    $i \ll^* j, z \ll^* t, z \dashrightarrow j \vDash i \ll^* t$;
Rule 5′    $: i \ll^* j, t \ll^* z, i \dashrightarrow z \vDash t \ll^* j$;
Rule 6    $i \dashrightarrow j, z \ll^* j, i \ll^* t \vDash z \ll^* t$;
Rule 6′    $: i \dashrightarrow j, i \ll^* z, t \ll^* j \vDash t \ll^* z$;

We see here the true role of constraints $i \dashrightarrow j$, which help us in inserting additional precedence constraints into the $Add_{\ll}$ set, with a strong impact on the antichain set $\mathcal{A}^*$ and on the optimal value of the related linear program $(\mathcal{P})_{Ant}$. Of course, any time such a pattern appears, it induces a Failure signal.

## 4.6 Branching Strategy

We described in Sect. 4.2 the Branching mechanism, which relies on the extraction of some forbidden pattern $i_1, j_1, i_2, j_2$:

- $i_1 \ll j_1$ and $i_2 \ll j_2$; (C5)
- no other pair $u \ll v$ exists with $u, v \in \{i_1, i_2, j_2, j_1\}$ than the pairs in (C5) above.

Since the way branching parameters are chosen is a critical issue as soon as Branch/bound and constraint propagation are performed. So we must now specify the strategy which is used here in order to compute a well-fitted 4-uple $i_1, j_1, i_2, j_2$. As a matter of fact, we apply here the well-known "most constraint variable" principle, and focus on the shortest circuits of the subgraph $(\mathcal{A}(z^*), E_{\ll})$ and on the antichains in $\mathcal{A}(z^*)$ which are the most involved in those circuits. As told in Sect. 4.4, branching has to be performed only if there exists some circuit in the subgraph $(\mathcal{A}(z^*), E_{\ll})$, where $z^*$ is the optimal solution of the LP $(\mathcal{P})_{Ant}$, solved after constraint propagation has been performed. Then we distinguish two cases:

- **First case**: there exists a circuit with length 2. In such a case, circuits with length 2 define in a natural way a non oriented graph $(\mathcal{A}(z^*), F)$ on the set $\mathcal{A}(z^*)$: two antichains $a, a'$ in $\mathcal{A}(z^*)$ define an edge of this graph if they also define a circuit of the oriented graph $(\mathcal{A}(z^*), E_{\ll})$. We consider an antichain $a_0$ which is with maximal degree $D_F(a_0)$ in the graph $(\mathcal{A}(z^*), F)$, together with some antichain $a_1$, with maximal degree $D_F(a_1)$ among the $F$-neighbours of $a_0$. Then we derive the forbidden pattern $i_1, j_1, i_2, j_2$, according to the proof of Theorem 2 in Sect. 3.4 and to Fig. 3a.
- **Second Case**: there does not exist any circuit with length 2. Then we compute the largest strongly connected component $A_0$ of the oriented graph $(\mathcal{A}(z^*), E_{\ll})$, together with the antichain $a_0$, which is such that:
  - There exists at least one pair $a_1, a_2$, such that $(a_1, a_0)$ and $(a_0, a_2)$ are in the arc set $E_{\ll}$, while $(a_1, a_2) \notin E_{\ll}$;
  - The sum $D_F^-(a_0) + D_F^+(a_0)$ of the inner and outer degrees of $a_0$ in the subgraph $(A_0, E_{\ll})$ induced from by $A_0$ is maximal.

Finally we compute some circuit $\Gamma$ which contains $a_0$ as well as $a_1, a_2$ above and which is with minimal length, and we derive the forbidden pattern $i_1, j_1, i_2, j_2$, according to the proof of Theorem 2 in Sect. 3.4 and to Fig. 3b.

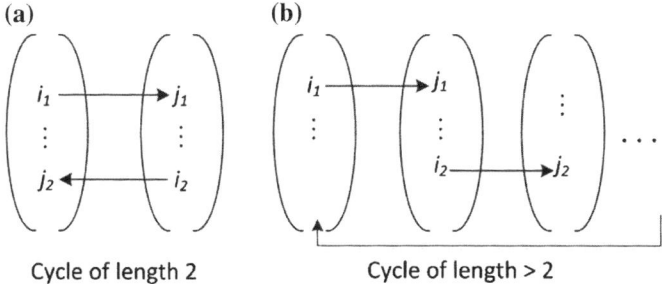

**Fig. 3** Extracting a forbidden pattern

## 4.7 Implementation

The global Branch/bound algorithm INT-ORD-ENUM Branch/Bound is implemented according to a Breadth First Search strategy which may be summarized as follows:

1. **Pretreatment**: Compute a feasible Non Preemptive RCPSP schedule $\sigma$, while using a greedy randomized insertion flow heuristic as in [21]. Derive an upper bound $UB$, together with an initial antichain subset $B \subset \mathcal{A}$, such that the linear program $(\mathcal{P})^B_{Ant}$ admits a feasible solution; Initialize the breadth search node list $L$ as the list $\{(Add_{\ll} = Nil, Add_{--\rightarrow} = Nil)\}$;

2. **Main Process: Breadth First Tree Search.** Let $L$ be the current node list, ordered according to LP $(\mathcal{P})^B_{Ant}$ related values, and $S$ be the first node in $L$; $S$ is defined by two additional constraint sets $Add_{\ll}$ and $Add_{--\rightarrow}$; Delete $S$ from $L$; Perform Constraint Propagation and extend $Add_{\ll}$ and $Add_{--\rightarrow}$; If *Failure* then go back to 2. Else go to 3.;

3. Solve the LP $(\mathcal{P})^B_{Ant}$ related to $S$ through column generation and test the oriented graph $(\mathcal{A}(z^*), E_{\ll})$ deriving from the obtained optimal solution $z^*$; If $1.z^* \geq UB$ then go to 2. Else go to 4.;

4. If the graph $(\mathcal{A}(z^*), E_{\ll})$ is acyclic then derive from $z^*$ a feasible schedule $\sigma$, update the upper global bound $UB$ and go back to 2. Else go to 5.;

5. Compute branching parameters $i_1, j_1, i_2, j_2$, according to Sect. 4.6 and create both related children:

    – 1 th son: insert $i_1 \ll j_2$ into the set $Add_{\ll}$;
    – 2 th son: insert $i_2 \ll j_1$ into the set $Add_{\ll}$; and $i_1 --\rightarrow j_2$ into the set $Add_{--\rightarrow}$;

    Insert those two children nodes in $L$, according to their related LP $(\mathcal{P})^B_{Ant}$ value; Go back to 2.;
    Process ends as soon as the LP value related to the first element of $S$ is no smaller than $UB$. Then current value $UB$ provides the optimal makespan value.

    This algorithm is implemented in C++, and linear programs $(\mathcal{P})^B_{Ant}$ and $L$-$PRICE(\lambda)$ are handled by CPLEX 12 linear solver. But the global INT-ORD-ENUM process is embedded into the SCIP framework for branch cut and price algorithms [22]. The SCIP framework consists in a template library which implements through breadth first search generic branch and bound schemes involving Linear Programming together with pricing scheme.

## 4.8 An Example

Let us consider an instance of 6 activities and 1 resource. Each activity has a duration equal to 1 and a resource requirement equal to 1. The precedence constraints are given by the precedence graph Fig. 4a. We initialize the set of antichains with the 6 singleton antichains. The tree constructed by our method and the branching decisions are given Fig. 5. The resulting optimal solution is given according to the Gantt chart Fig. 4b.

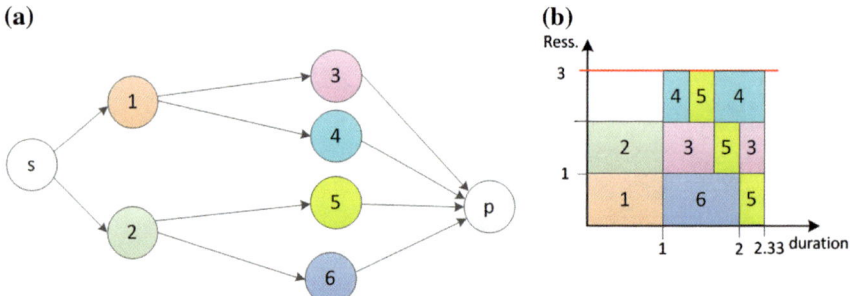

**Fig. 4** **a** Precedence graph. **b** Gantt chart of optimal preemptive solution

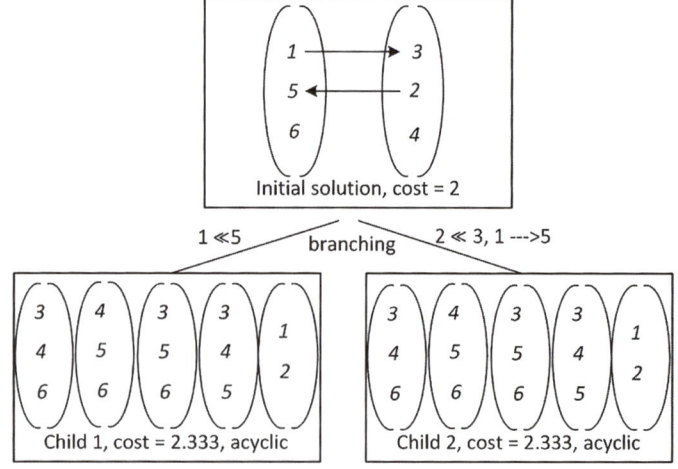

**Fig. 5** Solving instance of Fig. 4

# 5 Numerical Experiments

Experiments were carried on in C++, on linux CentOS proc. Intel(R) Xeon(R) 2.40 GHz. The instances which we used were PSPLIB instances [23].

Our main achievement (Sect. 5.1) here was to solve in an exact way and in a rather short time (never more that 95 CPU seconds) Preemptive RCPSP on all 30 activity instances of the PSPLIB library, which had been, until now, never done. Also, we could get an evaluation of the bounding process related to linear program $(\mathcal{P})_{Ant}^{B}$, and check that in average, $LB(I)$ approximates the optimal Non Preemptive RCPSP optimal value by less than 6%. By the same way, we checked that the augmented lower bound $LB^*(I)$ hardly improve $LB(I)$ by less than 0.5%. Finally (Sect. 5.2), though we were not able to handle in an exact way all 60/120 activity instances of the PSLIB library, we could derive new lower bounds for several Non Preemptive RCPSP instances of the PSPLIB library.

## 5.1 Exact Results on j30, Comparative Analysis

The columns of Table 1 have the following meaning:

- *No Preemp. opt.*: optimal value for Non Preemptive RCPSP (available in PSPLIB website)
- *Preemp. opt.*: optimal for preemptive RCPSP (our results)
- *#nodes*: number of nodes created
- *cpu (s)*: cpu time in seconds

Table 2 compares, for j30, $LB(I)$ and *Mean Prempt. Opt* values.

**Remark**: in almost 50% of the cases (exactly 236 instances among 480), the values $LB(I)$, *Premp. Op.* and *No Premp. Opt.* coincide.

**Table 1** Results on j30 of PSPLIB

|          | No Preemp. opt. | Preemp. opt. | #nodes  | cpu (s) |
| -------- | --------------- | ------------ | ------- | ------- |
| Mean     | 58.99           | 58.07        | 72.73   | 2.04    |
| Min      | 34.00           | 34.00        | 0.00    | < 0.01  |
| Max      | 129.00          | 129.00       | 2130.00 | 94.11   |
| Std dev. | 14.09           | 13.80        | 214.87  | 8.03    |

**Table 2** Evaluation of the bound $LB(I)$

| Mean $LB(I)$ | Mean $LB^*(I)$ | Mean Premp. opt. | Mean No Premp. opt. |
| ------------ | -------------- | ---------------- | ------------------- |
| 56.73        | 56.79          | 58.07            | 58.99               |

## 5.2 New Best Lower Bounds

Our method gives new best lower bound for j60, j90 and j120 instances (in a limit of time of 3 hours). The columns of Table 3 have the following meaning:

- *best No preemp. UB*: best known upper bound for no preemptive RCPSP (available in PSPLIB website)
- *Preemp. LB*: lower bound for preemptive RCPSP (our method)
- *Deduced no preemp. LB*: lower bound for no preemptive RCPSP which we deduce from *Preemp. LB*
- *Best known LB*: the best known lower bound currently available in PSPLIB website and updated with the recent results of [24].

# 6 Interval Order Based Heuristics for Preemptive RCPSP

## 6.1 Random Generation of Interval Orders

We first tried a very intuitive approach, which consists in randomly generating minimal interval order relations which are extensions of the $\ll$ precedence relation. In order to do it, we assigned every activity $i$ with some artificial auxiliary duration $v_i$, derived the related largest path schedule $\sigma(v)$ (which does not satisfy the resource constraints) and set, for any pair of activity $i, j$: $i \ll^* j$ iff $i$ precedes $j$ according to schedule $\sigma(v)$. We tested several strategies for the generation of duration vector $v$, among them:

- *Strategy S1*: $v = d$ (we keep on with the current duration vector $d$);
- *Strategy S2*: $v$ is generated in a full random way.

Then we got, on the j30 instances of PSPLIB the results Table 4.

**Comment:** this result may be viewed as rather surprising. It highlights both the complexity of the problem the fact, that, in most cases, the number of activities which are really preempted inside an optimal solution is not very high.

## 6.2 Heuristic "Minimal Activity First"

We propose now another heuristic approach, based upon the fact that, no activity may start before at least some *minimal* activity (in the sense of the $\ll$ relation) has been entirely performed. Telling this allows us to extend the *Antichain* linear program $(\mathcal{P})_{Ant}$ by introducing the following additional *Minimal-Activity-First* constraint:

$\sum_{a \text{ such that } a \subseteq Min \neq Nil} z_a \geq \min(d_i), i \in Min\text{-}Act$, where $Min\text{-}Act$ is the set $\{i \in X$, such that no activity $j$ exists such that $j \ll i\}$.

**Table 3** New best lower bounds

| Instance | Best non preemp. UB | Preemptive LB | Deduced non preemp. LB | Best known LB |
|----------|---------------------|---------------|------------------------|---------------|
| j6013_1.sm | 112 | 106.41 | 107 | 105 |
| j6029_2.sm | 133 | 126.20 | 127 | 123 |
| j6029_3.sm | 121 | 117.29 | 118 | 115 |
| j6029_4.sm | 134 | 129.29 | 130 | 126 |
| j6029_5.sm | 110 | 104.04 | 105 | 102 |
| j6029_6.sm | 154 | 145.30 | 146 | 144 |
| j6029_7.sm | 123 | 116.00 | 116 | 115 |
| j6029_9.sm | 112 | 106.83 | 107 | 105 |
| j6045_1.sm | 96 | 91.00 | 91 | 90 |
| j6045_2.sm | 144 | 137.32 | 138 | 134 |
| j6045_3.sm | 143 | 137.50 | 138 | 133 |
| j6045_4.sm | 108 | 102.49 | 103 | 101 |
| j6045_5.sm | 106 | 100.41 | 101 | 100 |
| j6045_6.sm | 144 | 136.42 | 137 | 132 |
| j6045_7.sm | 122 | 116.04 | 117 | 113 |
| j6045_8.sm | 129 | 122.17 | 123 | 119 |
| j6045_9.sm | 123 | 118.20 | 119 | 114 |
| j6045_10.sm | 114 | 106.48 | 107 | 104 |
| j9041_1.sm | 142 | 129.18 | 130 | 129 |
| j9045_3.sm | 154 | 144.43 | 145 | 144 |
| j9045_6.sm | 175 | 163.26 | 164 | 163 |
| j9045_8.sm | 160 | 150.26 | 151 | 150 |
| j9045_9.sm | 158 | 145.12 | 146 | 145 |
| j12036_4.sm | 236 | 217.35 | 218 | 217 |
| j12051_2.sm | 221 | 200.37 | 201 | 200 |
| j12051_5.sm | 230 | 205.88 | 206 | 205 |
| j12056_1.sm | 237 | 218.17 | 219 | 218 |
| j12056_3.sm | 241 | 222.12 | 223 | 220 |
| j12056_4.sm | 222 | 206.62 | 207 | 205 |
| j12056_5.sm | 280 | 261.80 | 262 | 261 |
| j12056_7.sm | 283 | 263.29 | 264 | 260 |
| j12056_8.sm | 289 | 268.04 | 269 | 265 |
| j12056_9.sm | 288 | 266.34 | 267 | 264 |

Let us call the *Augmented Antichain* LP the linear program $(Aug\text{-}\mathcal{P})_{Ant}$ which derives from $(\mathcal{P})_{Ant}$ through insertion of this additional constraint. It provides us with a lower bound *Aug-LB* of *Preemptive* RCPSP which is a little bit better than

**Table 4** Evaluation of the bound $LB(I)$

| Mean LB(I) | Mean Premp. opt. | Mean S1 | Mean S1 |
|---|---|---|---|
| 56.73 | 58.07 | 63.72 | 76.11 |

**Table 5** Minimal-Activity-First evaluation

| Mean LB(I) | Mean Aug. LB(I) | Min-Act-First | Random Min-Act-First |
|---|---|---|---|
| 56.73 | 56.79 | 60.51 | 60.34 |

the bound provide by $(\mathcal{P})_{Ant}$. But we see that we may enhance this lower bound, by setting, for any activity $i$ in *Min-Act*, the following constraint *MAF(i)*:

$MAF(i)$: $\sum_a$ such that $a \subseteq Min,$ and $i \in a \neq Nil$ $z_a \geq d_i, i \in Min\text{-}Act,$

and by denoting by $(Aug\text{-}\mathcal{P})^i_{Ant}$ the linear program which derives from $(\mathcal{P})_{Ant}$ by inserting the constraint *MAF(i)*. Clearly, the optimal value $W_i$ of $(Aug\text{-}\mathcal{P})^i_{Ant}$ is not a lower bound of *Preemptive* RCPSP, but, $W = \min_{i \in Min\text{-}Act} W_i$ is such a lower bound.

So we may try to apply to our *Preemptive* RCPSP problem the following scheme:

*Min-Act-First* Algorithmic Scheme:

**For** $p = 1..Card(X)$ **do**

Compute $W$ and $i_0$ such that $W = W_{i_0}$;

Let $z$ be the optimal solution of $(Aug\text{-}\mathcal{P})^{i_0}_{Ant}$;

Schedule all valid *antichains* of the set $Min(z, i) = \{a$ such that $a \subseteq Min,$ and $i \in a \neq Nil,$ and $z_a \neq 0\}$.

*Reduce* the activities $j \neq i$, which belong to some of these *antichains*, that means, for any such $j$, set: $d^*_j = d_j - \sum_{a \in Min(z,i)}$ such that $j \in a$ $z_a$.

Remove $i_0$ from $X$, and consider the RCPSP instance provided by the remaining activities provided with the $d^*$ durations.

As Table 5 shows, using the *Augmented Antichain* LP allows to significantly improve the quality of the lower bound. But the resulting *Min-Act-First* Algorithm happens to be rather efficient, and more if we randomize the process, by allowing it, at every iteration, to randomly choose $i_0$ as one of both activities which provides the smallest possible values $W_i$ (it misses the exact value of Preemptive RCPSP by less than 4 %).

## 7 Extension to Non Preemptive RCPSP

In *Non Preemptive* RCPSP, every activity must be performed as a whole without any interruption: for any activity $i \in X$, the temporal phase induced by the set of all instants $t$ when $i$ is running, define an interval $I_i$ of the time space. So, a schedule is entirely defined by the *starting-time* values $t_i, i \in X$. Then a natural reformulation of *Preemptive* RCPSP comes as follows:

**Theorem 3** *Solving the Non Preemptive RCPSP instance defined by $(X, K, \ll)$ means computing an extension $\ll^*$ of the $\ll$ precedence relation which is:*

- *an interval order, such that any $\ll^*$-antichain is valid;*
- *minimal, in the sense of inclusion, with those properties;*
- *such that the optimal value of related $(\mathcal{P})_{Ant}$ program is the smallest possible.*

*Proof* Left to the reader.

We are going to describe now in details the way our previous *Branch and Price* scheme may be adapted to the *Non Preemptive* case.

## 7.1 The Nodes of the Tree Search and the Bounding Process

In case when a current extension $\ll^*$ of $\ll$ is an interval order and admit some *non valid antichain a*, then, because of *Non Preemption*, we must impose that at least some pair $(i, j)$ of activities in $a$ to be such that $I_i \cap I_j = Nil$. We denote by $i \sim j$ this constraint $i \sim j \Leftrightarrow I_i \cap I_j = Nil \Leftrightarrow (i \ll^* j) \vee (j \ll^* i)$. That means that a node $\Sigma$ of the Tree Search process which we design here for *Non Preemptive* RCPSP is mainly defined by a collection $Add(\Sigma) = (Add_\ll, Add_{-\to}, Add_\sim)$ of additional constraints $i \ll^* j$, $i \dashrightarrow j$ and $i \sim j$, which we shall call the *master representation* of the node $\Sigma$. This master representation of $\Sigma$ gives rise, in a natural way, to auxiliary data $Add\text{-}Aux(\Sigma)$:

- A 2-dimensional vector *TAB*, with indexation on the set $X \cup \{Start, End\}$, where *Start* and *End* are two dummy activities with null durations and trivial semantics and with integral values: $TAB(i, j) = u$ means the constraint $t(j) \geq t(i) + u$. Clearly, this vector contains the information current upper bound UB as well as on the $\pi$, $\Pi$ values of Sect. 4: we should always have $TAB(End, Start) = -UB$ and, for any $i \in X$, $TAB(Start, i) = \pi(i)$ and $TAB(i, End) = \Pi(i)$;
- A binary symmetric relation *Int*, whose meaning is $i \ Int \ j \Leftrightarrow (i \dashrightarrow j) \wedge (j \dashrightarrow i)$;
- A clique family *CLI*, that means a collection $CLI(q), q = 1 \ldots N_{CLI}$ of subsets of $X$ such that for any pair $i, j$ in $CLI(q)$, we have $i \ Int \ j$. We provide any clique $CLI(q)$ with a lower bound $min(q)$ and an upper bound $max(q)$: the length of the interval $\cup_{I \in CLI(q)} I_x$ is at least equal to $min(q)$ and does not exceed $max(q)$.

*Bounding* is performed as in the preemptive case, that means through computation of the optimal value of the $(\mathcal{P})_{Ant}$ linear program related to the node $\Sigma$, with the difference that some additional constraints are inserted into $(\mathcal{P})_{Ant}$:

- *Non Intersection* Constraints: for any pair of activities $i, j$ such that $i \sim j$, which of course includes the pairs $i, j$ such that $i \ll^* j$ or such that $j \ll^* i$, we set: $\sum_{a \text{ such that } i, j \in a} z_a = 0$;
- *Clique* constraints: for any clique $CLI(q), q = 1 \ldots N_{CLI}$, $min(q) \leq \sum_{a \text{ such that } CLI(q) \subseteq a} z_a \leq max(q)$.

As a matter, of fact, we also insert into the $(\mathcal{P})_{Ant}$ program, any time a minimal (in the inclusion sense) non valid antichain $f$ is generated, the following constraints:

- *Minimal Non Valid Antichain* constraints: $\sum_{a \text{ such that } a\cap f\neq Nil} z_a \geq \delta(f) = $ sum of the two smallest durations of the tasks of $f$ (we do it of course only in the case $\delta(f) > Sup_{i\in f}d_i$), that means in the case when the optimal *makespan* of the *Non Preemptive* RCPSP instance defined by $f$ is equal to $\delta(f)$. This constraint expresses the fact that in any feasible schedule $\sigma$, the time which will be spent in running the tasks of $f$ will be at least equal to $\delta(f)$.

The pricing scheme which was used in Sect. 3.2 in order to deal with the $(\mathcal{P})_{Ant}$ program can be easily extended to the management of these new constraints, and experience shows that the induced additional computational cost remains small.

## *7.2 Propagation Rules*

The *TAB* vector is managed in such a way that, at any time, and for any triple $(i, j, k)$, we have $TAB(i, j) + TAB(j, k) \geq TAB(i, k)$ and also $TAB(i, i) = 0$. We keep all the rules which were described in Sect. 4. But we also introduce additional rules:

- Rules related to the *clique* collection CLI:

  - For every $i \in X$, we set $Q(i) = \{q = 1 \ldots N_{CLI}$ such that $i \in CLI(q)\}$; Every time a new relation $i$ $Int$ $j$ is inferred, we create the new cliques $\{i, j\}\cap(CLI(q)\cap CLI(r))$, $q \in Q(i)$, $r \in Q(j)$, while, of course, avoiding any duplication of a same clique. We derive related *min* and *max* values according to *ad hoc* simple computation;
  - For every $q = 1 \ldots N_{CLI}$, we keep memory of the set $EX(q)$ of tasks $i$ such that $CLI(q) \cup \{i\}$ is a non valid antichain. So, if $i \in EX(q)$, and if $i$ $Int$ $j$ for any $j$ in $CLI(q)$ but some task $j_0$, we infer $i \sim j_0$;
  - We set $q_1 \ll_{CLI} q_2$ if there exist $i \in CLI(q_1)$ and $j \in CLI(q_2)$ such that $i \ll^* j$, and, at any time during the process, we update the $\ll_{CLI}$ relation, and complete it in such a way it remains transitive. If $q$ is such that $q \ll_{CLI} q$ then we infer *Fail*; If $q_1, q_2$ are such that $q_1 \ll_{CLI} q_2$ then we infer, for any $i \in CLI(q_1)$ and $j \in CLI(q_2)$, that $j \dashrightarrow i$;
  - *Saturated Clique* Rule: If $q, r = 1 \ldots N_{CLI}, i \in EX(q) \cap CLI(r)$ are such that $q \ll_{CLI} r$, then we infer $q \ll_{CLI} r_1$, for any $r_1$ such $i \in CLI(r_1)$.

**Remark**: This last rule links the resource constraint with *Non Preemption*: if the period $P$ when actions of $CLI(q)$ are run cannot intersect the period $I_i$ when activity $i$ is run, then $P$ should be entirely before $I_i$ or after $I_i$. This is illustrated by following Fig. 6.

**Fig. 6** Forbidden pattern related to the *Saturated Clique* rule

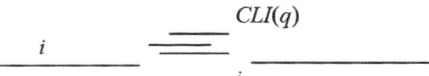

- Rules related to the *Non Intersection* $\sim$ relation:
  - In case $i, j, k$ are such that $i \lll^* j$, $i \dashrightarrow k$, and $k \sim j$ then we infer that $k \lll^* j$; clearly, we also may use a similar rules by changing the orientation of the $\lll^*$ and $\dashrightarrow$ relationships;
  - We prevent ourselves against the existence of any pattern $i, j, k, l \in X$, such that $i \sim j, k \sim l, i \ Int \ l, k \ Int \ j, k \ Int \ i, j \ Int \ l$;

  **Remark**: those two rules are related to the fact that we should be able to orient the *Non Intersection* $\sim$ relation in such a way it extends the $\lll^*$ relation into a partial order (transitive and anti-symetric) relation.

- We prevent ourselves against the existence of any pattern $i, j, k, l \in X$, such that $i \sim j, k \lll^* l, i \ Int \ k, j \ Int \ k, j \dashrightarrow l, k \dashrightarrow j$, or against the existence of any equivalent pattern modulo symmetry.

  **Remark:** this last rule expresses the fact that we should be able to orient the *Non Intersection* $\sim$ relation in such a way it extends the $\lll^*$ relation into an interval partial order, that means in a way which forbids the pattern of Fig. 1.

## 7.3 Closing a Node

Clearly, the occurrence of a *Fail* signal as an output of the constraint propagation process, or the fact that the lower bound $LB$ obtained by solving the linear program $(\mathcal{P})_{Ant}$ related to node $\Sigma$ is at least equal to current upper bound $UB$ closes the current node $\Sigma$. But there is also another situation which allows considering the problem induced by the constraint collection $Add(\Sigma)$ as a trivial problem. As a matter of fact, one sees that if 3 activities $i, j, k$ are such that $i \sim j, i \sim k$ and $j \ Int \ k$, then orienting the disjunction $i \sim j$ according to the relation $i \lll^* j$ is equivalent to orienting the disjunction $i \sim k$ according to the relation $i \lll^* k$. It comes that for any pair $(i, j)$ such that $i \sim j$, we may set $Eq(i, j) = \{$the set of all oriented pairs $(k, l), k \sim l$, such that $k \lll^* l$ derives from $i \lll^* j$ through propagation of the above mechanism$\}$. Of course, if it happens that, for some pair $(i, j), i \sim j$, we get that $(j, i) \in Eq(i, j)$, then we also get a *Fail* signal. Performing this consistency test in any node $\Sigma$ of our tree search is not very easy to do. But, in case the collection $Add(\Sigma)$ is such that for any pair $i, j$ we know either $i \sim j$ or $i \ Int \ j$, then we may state:

**Theorem 4** *In case the collection $Add(\Sigma)$ is such that for any pair $i, j$ we know either $i \sim j$ or $i \ Int \ j$, then it is possible to extend the $\lll^*$ as an interval order if and* *only if we never have* $(j, i) \in Eq(i, j)$. *In such a case, the makespan related to any such extension is always equal to the optimal value of the* $(\mathcal{P})_{Ant}$ *linear program.*

*Proof* Can be deduced from the general characterization of interval graphs in [25].

It comes that, in case $Add(\Sigma)$ is full, that means is such that for any pair $i, j$ we know either $i \sim j$ or $i$ *Int* $j$, then we close the node by building the equivalence classes $Eq(i, j)$ and testing that no arc $(i, j)$ is equivalent to its opposite arc $(j, i)$ in the sense of those classes. In case there exist $i, j$ such that $(j, i) \in Eq(i, j)$ then we close the node with a *Fail* signal, else, we do it because $UB$ becomes equal to current $LB$.

## 7.4 Updating the Upper Bound UB

Just before performing the branching process, we try to compute, in a greedy way, a feasible solution of the RCPSP extension induced by current collection $Add(\Sigma)$. Our heuristic greedy procedure *Greedy-Schedule* works this way:

- We derive from the *TAB* vector *time windows* for the starting-time $t_i, i \in X$;
- So let us consider that we have been already scheduling some activity subset $Y \subset X$, in such a way that if $i \in Y$, then any activity $j$ such that $t_j$ needs to be smaller than $t_i$ according to those *time windows*, is also in the set $Y$. Then, for any task $i \in X - Y$, such that no $j \in X - Y$ exists such that $t_j$ should be smaller than $t_i$, we compute the smallest starting-time $\tau(i)$ such that we may schedule $i$ at starting-time $\tau(i)$ without violating any constraint. Of course, it may occur that $\tau(i)$ does not exist, and this makes *Greedy-Schedule* stop;
- We define a task $i \in X - Y$ as being *eligible* if no activity $j$ exists in $X - Y$ such that $t_j$ should be smaller than $t_i$ or $\tau(j) + d_j \leq \tau(i)$. Then we pick up an *eligible* activity $i$ and schedule it at starting-time $\tau(i)$.

As a matter of fact, we notice that this process may be turned non deterministic, and cast into a Monte-Carlo replication scheme, while performing $N$ replications of the above greedy heuristic, and keeping with the best schedule it ever produced.

## 7.5 Branching

Scheduling the activities of $X$ by setting, for any task $i \in X$, Starting-time of $i = t_i = TAB(Start, i)$ provides us with a schedule $\sigma$ which is consistent with all constraints but the disjunctive constraints related to resources. In case, this schedule $\sigma$ does not satisfy those resource constraints, it makes appear *non valid antichains*. So we select such a *non valid antichain* $a$, and branch as follows:

- In case $a$ involves no more than 3 activities, we consider all options $i \ll^* j$, with $i, j \in a$: we do it while giving priority to the oriented pairs $(i, j)$ according to $t_i + d_i - t_j$ increasing values.
- In case $a$ involves more than 3 activities, we pick up two tasks $i, j$ in $a$ and branch according to the options: $i \ll^* j, j \ll^* i, i \ Int \ j$.

*Branching Strategy:* We choose $a$ with minimal number of activities ($Card(a)$ is the smallest possible). In case of tie, we impose $a$ to be such that the length of the period during which the activities of $a$ are simultaneously running is the largest possible.

## 7.6 Numerical Test

While using the same hardware and software as in Sect. 5, we are currently implementing and testing this algorithm on the 480, 30-activity instances of PSPLIB, successfully solving, in less than 10 min, more than 400 instances while only using part of the propagation rules which we mentioned above.

## 8 Conclusion

In case of *Preemptive* RCPSP, our *Branch and Price* method is both very efficient and innovative, since we are the first to have ever solved all 30 activity instances of PSPLIB. Still, we need to find ways to improve it, since, at current time, we are not able to close all 60 activity instances. The complexity of the problem make the error induced by heuristics rather significant. Forthcoming research should be about the extension of our approach to *Non Preemptive* RCPSP.

## References

1. Kolisch, R., Padman, R.: Deterministic project scheduling. Omega **48**, 249–272 (1999)
2. Brucker, P., Drexl, A., Mohring, R., Neumann, K., Pesch, E.: Resource-constrained project scheduling: notation, classification, models and methods. EJOR **112**, 3–41 (1999)
3. Herroelen, W.: Project scheduling–theory and practice. Prod. Oper. Manage. **14**(4), 413–432 (2006)
4. Liu, S.S., Wang, C.J.: RCPSP profit max with cash flow. Aut. Const. **17**, 966–74 (2008)
5. Baptiste, P., Lepape, C.L.: Constraint propagation and decomposition techniques for highly disjunctive and highly cumulative project scheduling problems. Constraints **5**(1/2), 119–139 (2000)
6. Brucker, P., Knust, S.: A linear programming and constraint propagation-based lower bound for the RCPSP. EJOR **127**(2), 355–362 (2000)
7. Hartmann, S., Briskorn, D.: A survey of variants of RCPSP. EJOR **207**, 1–14 (2010)

8. Orji, M.J., Wei, S.: Project scheduling under resource constraints: a recent survey. Int. J. Eng. Res. Technol. (IJERT) **2**(2), 102–127 (2013)
9. Herroelen, W., Leus, R.: Project scheduling under uncertainty: survey and research potentials. EJOR **165**(2), 289–306 (2005)
10. Demeulemeester, E., Herroelen, W.: An efficient optimal solution procedure for the preemptive resource-constrained project scheduling problem. EJOR **90**, 334–348 (1996)
11. Verma, S.: Exact methods for the preemptive resource-constrained project scheduling problem, research and publication. Indian Institute of Management, Ahmedabad, India, 8 March 2006.
12. Nadjafi, B.A., Shadrokh, S.: The preemptive resource-constrained project scheduling problem subject to due dates and preemption penalties. J. Ind. Eng. **1**, 35–39 (2008)
13. Ballestin, F., Valls, V., Quintanilla, S.: Preemption in resource-constrained project scheduling. EJOR **189**, 1136–1152 (2008)
14. Vanhoucke, M., Debels, D.: Impact of various activity assumptions on the lead time and resource utilization of resource-constrained projects. Comput. Indust. Eng. **54**, 140–154J (2008)
15. Vanhoucke, M.: A genetic algorithm for the net present value maximization for resource constrained projects. In: Cotta, C., Cowling, P. (eds.) Evolutionary Computation in Combinatorial Optimization. Lecture Notes in Computer Science, vol. 5482, pp. 13–24. Springer, Berlin, Heidelberg (2009)
16. Peteghem, V.V., Vanhoucke, M.: A genetic algorithm for the pre-emptive and non pre-emptive muti-mode resource constraint project scheduling problem. EJOR **201**(2), 409–418 (2010)
17. Mingozzi, A., Maniezzo, V., Ricciardelli, S., Bianco, L.: Exact algorithm for RCPSP based on a new mathematical formulation. Manage. Sci. **44**, 714–729 (1998)
18. Carlier, J., Neron, E.: On linear lower bounds for the resource constrained project scheduling problem. EJOR **149**(2), 314–324 (2003)
19. Damay, A., Quilliot, A., Sanlaville, E.: Linear programming based algorithms for preemptive and non preemptive RCPSP. EJOR **182**(3), 1012–1022 (2007)
20. Roberts, F.S.: Discrete Maths Models. Prentice Hall, Englewood Cliffs (1976)
21. Quilliot, A., Toussaint, H.: Flow polyedra and RCPSP. RAIRO-RO **46**(4), 379–409 (2012)
22. http://scip.zib.de/
23. http://webserver.wi.tum.de/psplib/
24. Schutt, A., Feydy, T., Stuckey, P.J.: Explaining Time-Table-Edge-Finding propagation for the cumulative resource constraint. In: Gomes, C., Sellmann, M. (eds.) Integration of AI and OR Techniques in Constraint Programming for Combinatorial Optimization Problems. Lecture Notes in Computer Science, pp. 234–250. Springer, Berlin (2013)
25. Golumbic, M.C.: Algorithmic Graph Theory and Perfect Graphs. Academic Press, New York (1980)

# Population Size Influence on the Genetic and Ant Algorithms Performance in Case of Cultivation Process Modeling

Olympia Roeva, Stefka Fidanova and Marcin Paprzycki

**Abstract** In this paper, an investigation of the influence of the population size on the Genetic Algorithm (GA) and Ant Colony Optimization (ACO) performance for a model parameter identification problem, is considered. The mathematical model of an *E. coli* fed-batch cultivation process is studied. The three model parameters—maximum specific growth rate ($\mu_{max}$), saturation constant ($k_S$) and yield coefficient ($Y_{S/X}$) are estimated using different population sizes. Population sizes between 5 and 200 chromosomes and 5 and 100 ants in the population are tested with constant number of generations. In order to obtain meaningful information about the influence of the population size a considerable number of independent runs of the GA are performed. The observed results show that the optimal population size is 100 chromosomes for GA and 70 ants for ACO for 200 generations. In this case accurate model parameters values are obtained in reasonable computational time. Further increase of the population size, above 100 chromosomes for GA and 70 ants for ACO, does not improve the solution accuracy. Moreover, the computational time is increased significantly.

**Keywords** Ant colony optimization · Genetic algorithm · Least square distance · Hausdorff distance

O. Roeva
Institute of Biophysics and Biomedical Engineering, Bulgarian Academy of Science, Acad. G. Bonchev Str., bl.105, 1113 Sofia, Bulgaria
e-mail: olympia@biomed.bas.bg

S. Fidanova (✉)
Institute of Information and Communication Technologies, Bulgarian Academy of Sciences, Acad. G. Bonchev Str., bl. 105, 1113 Sofia, Bulgaria
e-mail: stefka@parallel.bas.bg

M. Paprzycki
Systems Research Institute, Polish Academy of Sciences,
Warsaw and Management Academy, Warsaw, Poland
e-mail: marcin.paprzycki@ibspan.waw.pl

© Springer International Publishing Switzerland 2015
S. Fidanova (ed.), *Recent Advances in Computational Optimization*,
Studies in Computational Intelligence 580, DOI 10.1007/978-3-319-12631-9_7

107

# 1 Introduction

Metaheuristics, such as genetic algorithms and ant colony optimization, are widely used to solve various optimization problems [8, 13]. They are highly relevant for industrial applications, because they are capable of handling problems with non-linear constraints, multiple objectives, and dynamic components—properties that frequently appear in the real-world problems [16]. Since their introduction and sub-sequent popularization [17], the GA and ACO have been frequently used as an alter-native optimization tool to the conventional methods and have been successfully applied in a variety of areas, and still find increasing acceptance, for example:

- modelling and control of cultivation processes [7, 27, 28];
- model identification [1, 3, 11, 23], etc.

The metaheuristic algorithms require setting of the values of several algorithm components and parameters. These parameters values have great impact on perfor-mance and efficacy of the algorithm [14, 15, 22, 29]. Therefore, it is important to investigate the algorithm parameters influence on the performance of the developed metaheuristic algorithms. The aim is to find the optimal parameters values for the considered optimization problem. The optimal values for the parameters depend mainly on (i) the problem; (ii) the instance of the problem to deal with and (iii) the computational time that will be spend in solving the problem. Usually in the algorithm parameters tuning a compromise between solution quality and search time should be done.

For the parameter setting of metaheuristics, several automated approaches exist. These methods use (i) a single step of parameter tuning (prior to the practical use of the algorithm), or parameter control (self adaptation to the problem being optimized). Parameter control is well suited when one wants good average performances across diverse problems, but the needed computation overhead leads to less efficiency on specific problems, compared to parameter tuning [9]. Best known parameter tuning techniques are racing, sequential parameter optimization [5] and meta-parameter setting (sometimes referred as meta-algorithm [5]).

Population sizing has been one of the important topics to consider in evolutionary computation [2, 12, 30]. Various results about the appropriate population size can be found in the literature [25, 26]. Researchers usually argue that a "small" population size could guide the algorithm to poor solutions [18, 24, 30] and that a "large" population size could make the algorithm expend more computation time in finding a solution [18, 20, 21]. Due to significant influence of population size to the solution quality and search time [26] a more thorough research should be done for this GA parameter.

The main goal of this research is to carry out investigation of the influence of one of the key GA parameters—population size (number of chromosomes)—on the algorithm performance for identification of a cultivation process model. Parameter identification of non-linear cultivation process models is a hard combinatorial opti-mization problem for which exact algorithms or traditional numerical methods do not work efficiently. A non-linear mathematical model of fed-batch cultivation process of the most important host organism for recombinant protein production—bacteria *E. coli*—is considered [26].

The paper is organized as follows. The problem formulation is given in Sect. 2. The GA and ACO algorithms are proposed in Sects. 3 and 4 respectively. The numerical results and a discussion are presented in Sect. 5. Conclusion remarks are done in Sect. 6.

## 2 Problem Formulation

### 2.1 E. Coli *Fed-batch Cultivation Model*

Application of the general state space dynamical model [6] for the *E. coli* cultivation fed-batch process leads to the following nonlinear differential equation system [26]:

$$\frac{dX}{dt} = \mu_{max} \frac{S}{k_S + S} X - \frac{F_{in}}{V} X \tag{1}$$

$$\frac{dS}{dt} = -\frac{1}{Y_{S/X}} \mu_{max} \frac{S}{k_S + S} X + \frac{F_{in}}{V} (S_{in} - S) \tag{2}$$

$$\frac{dV}{dt} = F_{in} \tag{3}$$

where $X$ is the biomass concentration, [g/l]; $S$ is the substrate concentration, [g/l]; $F_{in}$ is the feeding rate, [l/h]; $V$ is the bioreactor volume, [l]; $S_{in}$ is the substrate concentration in the feeding solution, [g/l]; $\mu_{max}$ is the maximum value of the specific growth rate, [$h^{-1}$]; $k_S$ is the saturation constant, [g/l]; $Y_{S/X}$ is the yield coefficient, [−].

The initial process conditions are [4]:

- $t_0 = 6.68$ h,
- $X(t_0) = 1.25$ g/l and $S(t_0) = 0.8$ g/l,
- $S_{in} = 100$ g/l.

For the considered non-linear mathematical model of *E. coli* fed-batch cultivation process the parameters that should be identified are:

- maximum specific growth rate ($\mu_{max}$),
- saturation constant ($k_S$),
- yield coefficient ($Y_{S/X}$).

## 2.2 Optimization Criterion

In practical view, modelling studies are performed to identify simple and easy-to-use models that are suitable to support the engineering tasks of process optimization and, especially of control. The most appropriate model must satisfy the following conditions:

 (i) the model structure should be able to represent the measured data in a proper manner;
(ii) the model structure should be as simple as possible compatible with the first requirement.

The optimization criterion is a certain factor, whose value defines the quality of an estimated set of parameters. To evaluate the mishmash between experimental and model predicted data the Least Square Regression is used.

The objective consists of adjusting the parameters ($\mu_{max}$, $k_S$ and $Y_{S/X}$) of the non-linear mathematical model function (Eqs. 1–3) to best fit a data set. A simple data set consists of $n$ points (data pairs) $(x_i, y_i)$, $i = 1, 2, \ldots, n$, where $x_i$ is an independent variable and $y_i$ is a dependent variable whose value is found by observation. The model function has the form $f(x, \beta)$, where the $m$ adjustable parameters are held in the vector $\beta$, $\beta = [\mu_{max}\ k_S\ Y_{S/X}]$. The goal is to find the parameter values for the model which "best" fits the data. The least squares method finds its optimum when the sum $J$ of squared residuals:

$$J = \sum_{i=1}^{n} r_i^2$$

is a minimum. A residual is defined as the difference between the actual value of the dependent variable and the value predicted by the model. A data point may consist of more than one independent variable. For an example, when fitting a plane to a set of height measurements, the plane is a function of two independent variables, $x$ and $z$, say. In the most general case there may be one or more independent variables and one or more dependent variables at each data point.

$$r_i = y_i - f(x_i, \beta).$$

## 3 Genetic Algorithm

GA was developed to model adaptation processes mainly operating on binary strings and using a recombination operator with mutation as a background operator. The GA maintains a population of chromosomes, $P(t) = x_1^t, \ldots, x_n^t$ for generation $t$. Each chromosome represents a potential solution to the problem and is implemented as some data structure $Ch$. Each solution is evaluated to give some measure of its "fitness". Fitness of a chromosome is assigned proportionally to the value of the objective function of the chromosomes. Then, a new population (generation $t + 1$) is formed by selecting more fit chromosomes (selection step). Some members of the new population undergo transformations by means of "genetic" operators to form new solution. There are unary transformations $m_i$ (mutation type), which create new chromosomes by a small change in a single chromosome ($m_i : Ch \rightarrow Ch$), and higher order transformations $c_j$ (crossover type), which create new chromosomes by combining parts from several chromosomes ($c_j : Ch \times \cdots \times Ch \rightarrow Ch$). After some number of generations the algorithm converges—it is expected that the best chromosome represents a near-optimum (reasonable) solution. The combined effect of selection, crossover and mutation gives so-called reproductive scheme growth equation [16]:

$$\xi (Ch, t + 1) \geq$$

$$\xi (Ch, t) \cdot eval\, (Ch, t) / \bar{F}\, (t) \left[ 1 - p_c \cdot \frac{\delta\, (Ch)}{m - 1} - o\, (Ch) \cdot p_m \right]$$

The structure of the herewith used GA is shown by the pseudocode below (Fig. 1). Three model parameters are represented in the chromosome—$\mu_{max}, k_S$ and $Y_{S/X}$. The following upper and lower bounds of the model parameters are considered [28]:

$$0 < \mu_{max} < 0.7,$$

$$0 < k_S < 1,$$

$$0 < Y_{S/X} < 30.$$

Roulette wheel, developed by Holland [17] is the herewith used selection method. The probability, $p_i$, for each chromosome is defined by:

$$p[\text{Individual } i \text{ is chosen}] = \frac{F_i}{\sum_{j=1}^{PopSize} F_j}, \tag{4}$$

**Fig. 1** Pseudocode for GA

```
begin
    i = 0
    Initial population P(0)
    Evaluate P(0)
    while (not done) do
    (test for termination criterion)
    begin
        i = i + 1
        Select P(i) from P(i − 1)
        Recombine P(i)
        Mutate P(i)
        Evaluate P(i)
    end
end
```

where $F_i$ equals the fitness of chromosome $i$ and $PopSize$ is the population size.

To reproduce the chromosomes simple crossover and binary mutation according to [28] are applied. In proposed genetic algorithm fitness-based reinsertion (selection of offspring) is used.

For the considered here model parameter identification, the type of the basic operators in GA are as follows [28]:

- encoding—binary,
- fitness function—linear ranking,
- selection function—roulette wheel selection,
- crossover function—simple crossover,
- mutation function—binary mutation,
- reinsertion—fitness-based.

The values of GA parameters are [28]:

- generation gap, ggap = 0.97,
- crossover probability, xovr = 0.75,
- mutation probability, mutr = 0.01,
- maximum number of generations, maxgen = 200.

## 4 Ant Colony Optimization (ACO)

The ACO is a stochastic optimization method that imitates the behavior of real ants colonies. They manage to establish the shortest rout to nutrishment sources and back. Real ants foraging for food lay down quantities of pheromone (chemical cues) marking the path that they follow. An isolated ant moves essentially at random but an ant encountering a previously laid pheromone will detect it and decide to follow it with high probability and thereby reinforce it with a further quantity of pheromone. Thus if more the ants follow a trail, the more attractive that trail becomes. The original idea comes from observing the exploitation of food resources among ants, in which ants' individually limited cognitive abilities have collectively been able to find the shortest path between a food source and the nest.

The ACO is usually implemented as a team of intelligent agents, which simulate the ants behavior, walking around the graph representing the problem to solve, using mechanisms of cooperation and adaptation. The requirements of the ACO algorithm are as follows [8, 13]:

- The problem needs to be represented appropriately, which would allow the ants to incrementally update the solutions through the use of a probabilistic transition rules, based on the amount of pheromone in the trail and other problem specific knowledge.
- A problem-dependent heuristic function, that measures the quality of components that can be added to the current partial solution.

**Ant Colony Optimization**
Initialize number of ants;
Initialize the ACO parameters;
**while not** end-condition **do**
    **for** $k = 0$ **to** number of ants
        ant $k$ choses start node;
        **while** solution is not constructed **do**
            ant $k$ selects higher probability node;
        **end while**
    **end for**
    Update-pheromone-trails;
**end while**

**Fig. 2** Pseudocode for ACO

- A rule set for pheromone updating, which specifies how to modify the pheromone value.
- A probabilistic transition rule based on the value of the heuristic function and the pheromone value, that is used to iteratively construct a solution.

The structure of the ACO algorithm is shown by the pseudocode below (Fig. 2).

The transition probability $p_{i,j}$, to choose the node $j$ when the current node is $i$, is based on the heuristic information $\eta_{i,j}$ and the pheromone trail level $\tau_{i,j}$ of the move, where $i, j = 1, \ldots, n$.

$$p_{i,j} = \frac{\tau_{i,j}^a \eta_{i,j}^b}{\sum_{k \in Unused} \tau_{i,k}^a \eta_{i,k}^b}, \tag{5}$$

where $Unused$ is the set of unused nodes of the graph.

The higher the value of the pheromone and the heuristic information, the more profitable it is to select this move and resume the search. In the beginning, the initial pheromone level is set to a small positive constant value $\tau_0$; later, the ants update this value after completing the construction stage. The ACO algorithms adopt different criteria to update the pheromone level.

The pheromone trail update rule is given by:

$$\tau_{i,j} \leftarrow \rho\tau_{i,j} + \Delta\tau_{i,j}, \tag{6}$$

where $\rho$ models evaporation in the nature and $\Delta\tau_{i,j}$ is new added pheromone which is proportional to the quality of the solution. Thus better solutions will receive more pheromone than others and will be more desirable in a next iteration.

The values of ACO parameters in our application are:

- evaporation parameter $\rho = 0.5$,
- $a = b = 1$,
- number of generations = 200.

# 5 Numerical Results and Discussion

All computations are performed using a PC/Intel Core i5-2320 CPU @ 3.00GHz, 8 GB Memory (RAM), Windows 7 (64 bit) operating system and Matlab 7.5 environment.

A series of numerical experiments are performed to evaluate the influence of the population size in GAs and ACO on the accuracy of the obtained solution. Using mathematical model of the *E. coli* cultivation process (Eqs. 1–3) the model parameters—maximum specific growth rate ($\mu_{max}$), saturation constant ($k_S$) and yield coefficient ($Y_{S/X}$)—are estimated. For the identification procedures consistently different population sizes (from 5 to 200 chromosomes in the population for GA and from 5 to 100 ants for ACO) are used. The number of generations is fixed to 200. Because of the stochastic characteristics of the applied algorithms, series of 30 runs for each population size are performed.

In the Tables 1 and 2, obtained average, best and worst objective function values for considered population sizes of GA and ACO respectively, are presented. The results observed for computational time are listed in Tables 3 and 4 respectively.

The numerical experiments show that increasing the size of the population of 5–100 chromosomes significantly improves the resulting value of the objective function (average results)—from 6.1200 to 4.5406 (see Table 1). The further increase in the size of population (more than 100 chromosomes) does not lead to more accurate results. The subsequent increase in the population size leads only to an increase in computational time without improving the value of the objective function (average results)—from 26.8644 s (100 chromosomes) to 52.4782 s (200 chromosomes) versus. $J = 4.5406$ to $J = 4.5453$ (see Table 3).

**Table 1** GA algorithm performance for various population sizes—objective function

| Population size | Objective function $J$ | | |
| --- | --- | --- | --- |
| | Average | Best | Worst |
| 5 | 6.1200 | 4.8325 | 9.2958 |
| 10 | 5.8000 | 4.8548 | 9.6175 |
| 20 | 4.7660 | 4.4753 | 5.3634 |
| 30 | 4.6519 | 4.4816 | 5.0094 |
| 40 | 4.6359 | 4.4437 | 4.9669 |
| 50 | 4.6070 | 4.4488 | 4.8636 |
| 60 | 4.5886 | 4.4625 | 4.8013 |
| 70 | 4.5648 | 4.4384 | 4.7357 |
| 80 | 4.5782 | 4.4474 | 4.7463 |
| 90 | 4.5711 | 4.4496 | 4.7211 |
| 100 | 4.5406 | 4.4252 | 4.7017 |
| 110 | 4.5455 | 4.4332 | 4.7319 |
| 150 | 4.5511 | 4.4575 | 4.6717 |
| 200 | 4.5453 | 4.4359 | 4.7206 |

**Table 2** ACO algorithm performance for various population sizes—objective function

| Population size | Objective function $J$ | | |
|---|---|---|---|
| | Average | Best | Worst |
| 5 | 6.2523 | 5.0652 | 7.9011 |
| 10 | 6.0527 | 4.8083 | 8.0956 |
| 20 | 5.4330 | 4.9293 | 6.5924 |
| 30 | 5.2849 | 4.7408 | 6.2202 |
| 40 | 5.2853 | 4.8004 | 6.0784 |
| 50 | 5.2206 | 4.6598 | 4.1695 |
| 60 | 5.2184 | 4.8983 | 5.7759 |
| 70 | 5.1350 | 4.7739 | 5.6652 |
| 80 | 5.1324 | 4.8078 | 5.7891 |
| 90 | 5.1415 | 4.7856 | 5.6120 |
| 100 | 5.0885 | 4.8382 | 5.4866 |

**Table 3** GA algorithm performance for various population sizes—computational time

| Population size | Computational time (s) | | |
|---|---|---|---|
| | Average | Best | Worst |
| 5 | 4.9457 | 4.5552 | 5.6004 |
| 10 | 6.0039 | 5.6316 | 6.3648 |
| 20 | 7.6482 | 7.3008 | 7.9561 |
| 30 | 11.1115 | 10.8265 | 11.5129 |
| 40 | 12.9824 | 12.4957 | 13.3537 |
| 50 | 14.9087 | 14.3989 | 15.5377 |
| 60 | 17.2766 | 16.6141 | 20.3113 |
| 70 | 19.7601 | 19.1725 | 20.0617 |
| 80 | 22.1880 | 21.7153 | 22.6669 |
| 90 | 24.3414 | 23.9150 | 24.8198 |
| 100 | 26.8644 | 26.4890 | 27.8306 |
| 110 | 29.7057 | 29.1878 | 30.2642 |
| 150 | 39.7273 | 39.1407 | 40.3887 |
| 200 | 52.4782 | 51.3087 | 55.8952 |

Similar conclusions can be made for ACO algorithm performance regarding Tables 2 and 4. The numerical experiments show that increasing the population size from 5 to 70 ants improve the accuracy of the achieved average result—from $J = 6.2523$ to $J = 5.1350$. Further increase of the size of the population only increases the computational time without significant improvement of the results.

The best value of the objective function achieved by GA is similar to this achieved by ACO, but the average value achieved by GA is better. One generation performed

**Table 4** ACO algorithm performance for various population sizes—computational time

| Population size | Computational time (s) | | |
|---|---|---|---|
| | Average | Best | Worst |
| 5 | 16.8065 | 16.5673 | 17.0509 |
| 10 | 29.5950 | 29.3126 | 29.8274 |
| 20 | 55.0699 | 54.1323 | 56.7376 |
| 30 | 90.1941 | 88.9674 | 91.1202 |
| 40 | 111.2729 | 109.2163 | 116.0803 |
| 50 | 131.8193 | 131.0720 | 133.4745 |
| 60 | 151.7526 | 148.8406 | 159.7606 |
| 70 | 173.8225 | 172.5839 | 177.0143 |
| 80 | 197.8873 | 196.5457 | 199.5877 |
| 90 | 234.9069 | 232.1607 | 238.2759 |
| 100 | 260.4468 | 258.3689 | 268.0097 |

by ACO is much slower than one generation performed by GA, thus the GA computational time is less than ACO computational time. We can conclude that GA performs better than ACO for this problem and the best population size for GA is 100 chromosomes.

For better interpretation the obtained numerical results are graphically visualized in the next figures. On Fig. 3 the objective function values, obtained during the 30 GA runs for 5, 10, 20 and 30 chromosomes in the population, are shown. The graphical results show that the GA could not find accurate solution using small population size—5 or 10 chromosomes. It needs at least 20 chromosomes in population for achieving a better solution. On Fig. 4 the objective function values, obtained during the 30 algorithm runs for 100, 110, 150 and 200 chromosomes in the population, are shown. Here, it could be seen that using large population size (110, 150 or 200 chromosomes) did not result in an improvement of the objective function values.

The ANOVA test is applied and the values of the objective function for population size equal and more than 100 are statistically equal. Moreover, as can be seen from Fig. 6 increasing the population size result in an acceleration of computational time. When the population size increases it leads to increase of the needed computational resources like time and memory which can be a problem for large-scale tests. Therefore we can conclude that populations with 100 individuals is optimal with respect to the value of the objective function and the needed computational resources.

All numerical experiments for the influence of the population size on the objective function value and on the computational time are summarized in Figs. 5 and 6. It can be concluded that for the considered here non-linear cultivation model parameter identification problem the optimal population size is 100 chromosomes in the population (for 200 generations).

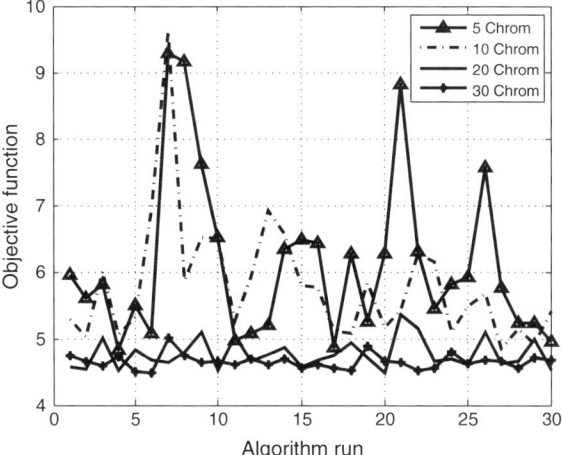

**Fig. 3** Objective function values obtained during the 30 algorithm runs for 5, 10, 20 and 30 chromosomes in the population

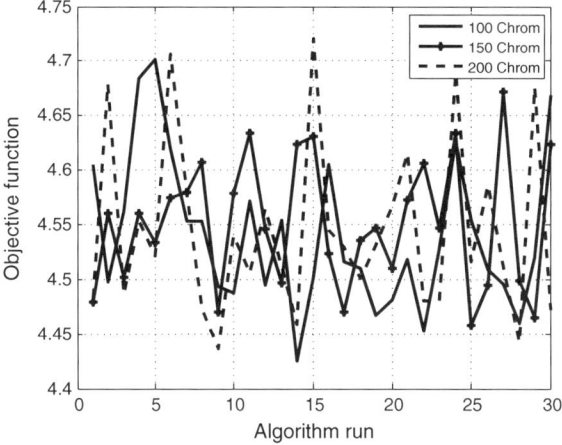

**Fig. 4** Objective function values obtained during the 30 algorithm runs for 100, 150 and 200 chromosomes in the population

In Table 5 the best parameter values ($\mu_{max}$, $k_S$ and $Y_{S/X}$), obtained using GA with 100 chromosomes in the population, are presented. According to [10, 19, 31] the values of the estimated model parameters are in admissible boundaries.

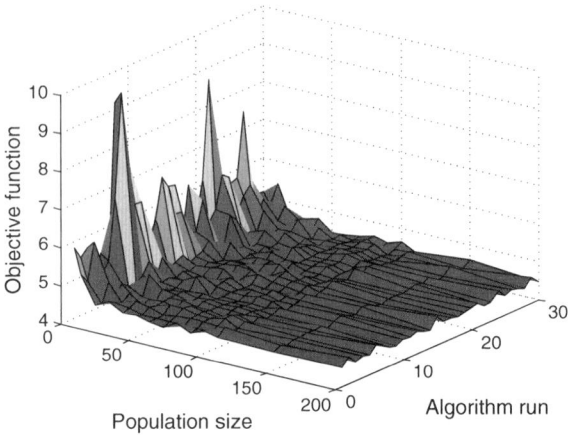

**Fig. 5** Influence of the population size on the objective function value

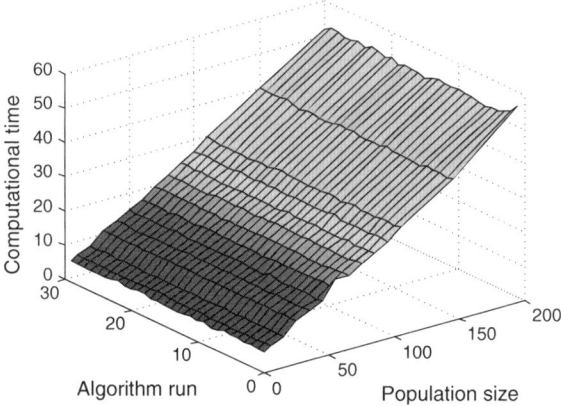

**Fig. 6** Influence of the population size on the computational time

**Table 5** Best parameter values of the model (100 chromosomes)

| Parameter | Value |
|---|---|
| $\mu_{max}$, (1/h) | 0.4881 |
| $k_S$, (g/l) | 0.0120 |
| $Y_{S/X}$ $(-)$ | 2.0193 |

# 6 Conclusion

A good selection of the algorithm parameters improve both computation time and solution accuracy. Finding good parameter values is not a trivial task and requires human expertise as well as time. In this paper, the influence of one of the key GA and ACO parameters (population size) on the algorithm performance, is studied. As a test problem, the *E. coli* fed-batch cultivation model parameter identification, is considered. The three model parameters (maximum specific growth rate ($\mu_{max}$), saturation constant ($k_S$) and yield coefficient ($Y_{S/X}$)) are identified. For a fixed number of the generations (200) different population sizes of the GA and ACO are explored. The numerical experiments are started with 5 chromosomes or ants in the population and consistently increased to 200 chromosomes for GA and 100 ants for ACO. The obtained results show that the optimal population size, for the GA considered here case study, is 100 chromosomes and 70 ants for ACO. Thus, accurate model parameters values are obtained with reasonable computational efforts. The use of smaller populations result in lower accuracy of the solution, obtained for a smaller computational time. The further increase of the population size increases the accuracy of solution. This effect is observed to a population size of 100 chromosomes for GA and 70 ants for ACO. The use of larger populations does not improve the solution accuracy and only increase the needed computational resources. The GA algorithm performs better than ACO for this application. It is faster and achieves better average value of objective function.

**Acknowledgments** Work presented here is a part of the Poland-Bulgarian collaborative Grant "Parallel and distributed computing practices" and by European Commission project ACOMIN.

# References

1. Akpinar, S., Bayhan, G.M.: A hybrid genetic algorithm for mixed model assembly line balancing problem with parallel workstations and zoning constraints. Eng. Appl. Artif. Intell. **24**(3), 449–457 (2011)
2. Alander, J.T.: On optimal population size of genetic algorithms. In: Proceedings of the IEEE Computer Systems and Software Engineering, p. 6569 (1992)
3. Al-Duwaish, H.N.: A genetic approach to the identification of linear dynamical systems with static nonlinearities. Int. J. Syst. Sci. **31**(3), 307–313 (2000)
4. Arndt, M., Hitzmann, B.: Feed Forward/feedback Control of Glucose Concentration during Cultivation of *Escherichia coli*. In: 8th IFAC Int, Conference on Comparative Applications in Biotechnology, pp. 425–429, Canada (2001)
5. Bartz-Beielstein, T.: Experimental Research in Evolutionary Computation: The New Experimentalism. Natural Computing Series. Springer, Berlin (2006)
6. Bastin, G., Dochain, D.: On-Line Estimation and Adaptive Control of Bioreactors. Elsevier, Amsterdam (1991)
7. Benjamin, K.K., Ammanuel, A.N., David, A., Benjamin, Y.K.: Genetic algorithm using for a batch fermentation process identification. J. Appl. Sci. **8**(12), 2272–2278 (2008)
8. Bonabeau, E., Dorigo, M., Theraulaz, G.: Swarm Intelligence: From Natural to Artificial Systems. Oxford University Press, New York (1999)
9. Clune, J., Goings, S., Punch, B., Goodman, E.: Investigations in meta-gas: panaceas or pipe dreams. In: GECCO 05 Proceedings, pp. 235–241 (2005)
10. Contiero, J., Beatty, C., Kumari, S., DeSanti, C.I., Strohl, W.L.; WolfeA.: Effects of mutations in acetate metabolism on high-cell-density growth of *Escherichia coli*. J. Ind. Microbiol. Biotechnol. **24**, 421–430 (2000)
11. da Silva, M.F.J., Perez, J.M.S., Pulido, J.A.G., Rodriguez, M.A.V.: AlineaGA—A genetic algorithm with local search optimization for multiple sequence alignment. Appl. Intell. **32**, 164–172 (2010)

12. Diaz-Gomez, P.A., Hougen, D. F.: Initial population for genetic algorithms. In: Hamid R. Arabnia., Jack Y. Yang., Mary Qu Yang. (eds.) A Metric Approacs, Proceedings of the International Conference on Genetic and Evolutionary Methods, GEM 2007, pp. 43–49. Las Vegas, Nevada, USA (2007)
13. Dorigo, M., Stutzle, T.: Ant Colony Optimization. MIT Press, London (2004)
14. Eiben Á, E., Hinterding, R., Michalewicz, Z.: Parameter control in evolutionary algorithms. IEEE Trans. Evol. Comput. **3**(2), 124–141 (1999)
15. Fidanova, S.: Simulated annealing: A monte carlo method for gps surveying. Comput. Sci. Lect. Notes Comput. Sci. **3991**, 1009–1012 (2006)
16. Goldberg, D.E.: Genetic Algorithms in Search. Optimization and Machine Learning. Addison Wesley Longman, London (2006)
17. Holland, J.H.: Adaptation in Natural and Artificial Systems, 2nd edn. MIT Press, Cambridge (1992)
18. Koumousis, V.K., Katsaras, C.P.: A sawtooth genetic algorithm combining the effects of variable population size and reinitialization to enhance performance. IEEE Trans. Evol. Comput. **10**(1), 19–28 (2006)
19. Levisauskas, D., Galvanauskas, V., Henrich, S., Wilhelm, K., Volk, N., Lubbert, A.: Model-based optimization of viral capsid protein production in fed-batch culture of recombinant *Escherichia coli*. Bioprocess Biosyst. Eng. **25**, 255–262 (2003)
20. Lobo, F.G., Goldberg, D.E.: The parameterless genetic algorithm in practice. Inf. Sci. Inform. Comput. Sci. **167**(1–4), 217–232 (2004)
21. Lobo, F.G., Lima, C.F.: A review of adaptive population sizing schemes in genetic algorithms. In: Proceedings of the Genetic and Evolutionary Computation Conference, pp. 228–234 (2005)
22. Nowotniak, R., Kucharski, J.: GPU-based Tuning of Quantum-Inspired Genetic Algorithm for a Combinatorial Optimization Problem. In: Proceedings of the XIV International Conference System Modeling and Control, ISSN 978–83-927875-1-8 (2011)
23. Paplinski, J.P.: The genetic algorithm with simplex crossover for identification of time delays, Intelligent, Information Systems, pp. 337–346 (2010)
24. Kumar, S.M., Jain, R., Anantharaman, N., Dharmalingam, V., Sheriffa-Begum, K.M.M.: Genetic algorithm based PID controller tuning for a model bioreactor. J. Indian Chem. Eng., **50**(3), 214–226 (2008)
25. Reeves, C.R.: Using genetic algorithms with small populations. In: Proceedings of the Fifth International Conference on Genetic Algorithms, pp. 92–99. San Francisco (1993)
26. Roeva, O.: Improvement of genetic algorithm performance for identification of cultivation process models. In: Advanced Topics on Evolutionary Computing. Artificial Intelligence Series-WSEAS, pp. 34–39 (2008)
27. Roeva, O., Slavov, T.S.: Fed-batch Cultivation Control based on Genetic Algorithm PID Controller Tuning. Lecture Notes on Computer Science. Springer, Berlin (2011)
28. Roeva, Fidanova, S.: A comparison of genetic algorithms and ant colony optimization for modeling of *E. coli* cultivation process. In: Real-World Application of Genetic Algorithms, pp. 261–282. In Tech. (2012)
29. Saremi, A., ElMekkawy, T.Y., Wang, G.G.: Tuning the parameters of a memetic algorithm to solve vehicle routing problem with backhauls using design of experiments. Int. J. Oper. Res. **4**(4), 206–219 (2007)
30. Piszcz, A., Soule, T.: Genetic programming: optimal population sizes for varying complexity problems. In: Proceedings of the Genetic and Evolutionary Computation Conference, pp. 953–954 (2006)
31. Zelic, B., Vasic-Racki, D., Wandrey, C., Takors, R.: Modeling of the pyruvate production with *Escherichia coli* in a fed-batch bioreactor. Bioprocess Biosyst. Eng. **26**, 249–258 (2004)

# A Hybrid Approach to Modeling, Solving and Optimization of the Constrained Decision Problems

Pawel Sitek and Jarosaw Wikarek

**Abstract** The paper presents a concept and implementation of a novel hybrid approach to the modelling, solving and optimization of the constrained decision problems. Two environments, mathematical programming (MP) and constraint programming (CP), in which constraints are treated in different ways and different methods are implemented, were combined to use the strengths of both. This integration and hybridization, complemented with an adequate transformation of the problem, facilitates a significant reduction of the combinatorial problem. The whole process takes place at the implementation layer, which makes it possible to use the structure of the problem being solved, implementation environments and the very data. The superiority of the proposed approach over the classical scheme is proved by 1/considerably shorter search time and 2/example-illustrated wide-ranging possibility of expanding the decision and/or optimization models through the introduction of new logical constraints, frequently encountered in practice. The proposed approach is particularly important for the decision models with an objective function and many discrete decision variables added up in multiple constraints. To validate the proposed approach, two illustrative examples are presented and solved. The first example is the authors' original model of cost optimization in the supply chain with multimodal transportation. The second one is the two-echelon variant of the well-known Capacitated Vehicle Routing Problem, 2E-CVRP. Distance, this metric is more realistic for the considered problem.

**Keywords** Discrete optimization · Constraint programming · Mathematical programming · Decision supporte

P. Sitek (✉) · J. Wikarek
Institute of Management and Control Systems,
University of Technology Kielce, Kielce, Poland
e-mail: sitek@tu.kielce.pl

J. Wikarek
e-mail: j.wikarek@tu.kielce.pl

© Springer International Publishing Switzerland 2015

121

S. Fidanova (ed.), *Recent Advances in Computational Optimization*,
Studies in Computational Intelligence 580, DOI 10.1007/978-3-319-12631-9_8

# 1 Introduction

The most of the models [1–5] of decision support and/or optimization in manufacturing, distribution, supply chain management, etc., have been formulated as the mixed integer linear programming (MILP) or integer programming (IP) problems and solved using the operations research (OR) methods. Their structures are similar and proceed from the principles and requirements of mathematical programming. The constraint-based environments have the advantage over traditional methods of mathematical modeling in that they work with a much broader variety of interrelated constraints (resource, time, technological, and financial) and allow producing "natural" solutions for highly combinatorial problems.

## 1.1 Constraint-Based Environments

We strongly believe that the constraint-based environment [6–8] offers a very good framework for representing the knowledge and information needed for the decision support. The central issue for a constraint-based environment is a constraint satisfaction problem. Constraint satisfaction problem (CSP) is a mathematical problem defined as a set of elements whose state must satisfy a number of constraints. CSP represent the entities in a problem as a homogeneous collection of finite constraints over variables, which are solved using constraint satisfaction methods. CSPs are the subject of intense study in both artificial intelligence and operations research, since the regularity in their formulation provides a common basis for analyzing and solving the problems of many unrelated families [6]. Formally, a constraint satisfaction problem is defined as a triple $(X,D,C)$, where $X$ is a set of variables, $D$ is a domain of values, and $C$ is a set of constraints. Every constraint is in turn a pair $(r, R)$ (usually represented as a matrix), where $r$ is an n-tuple of variables and $R$ is an n-ary relation on $D$. An evaluation of the variables is a function from the set of variables to the domain of values, $v:X \rightarrow D$. An evaluation $v$ satisfies constraint $((x_1, \ldots, x_n), R) if (v(x_1), \ldots, v(x_n)) \in R$. A solution is an evaluation that satisfies all constraints.

Constraint satisfaction problems on finite domains are typically solved using a form of search. The most widely used techniques include variants of backtracking, constraint propagation, and local search. Constraint propagation embeds any reasoning that consists in explicitly forbidding values or combinations of values for some variables of a problem because a given subset of its constraints cannot be satisfied otherwise [27]. CSPs are frequently used in constraint programming. Constraint programming is the use of constraints as a programming language to encode and solve problems. Constraint logic programming (CLP) is a form of constraint programming (CP), in which logic programming is extended to include concepts from constraint satisfaction. A constraint logic program is a logic program that contains constraints in the body of clauses. The CP and CLP environments are declarative.

The declarative approach and the use of logic programming provide incomparably greater possibilities for decision problems modeling than the pervasive approach based on mathematical programming.

## 1.2 Paper Contents

In this paper we focus on the problem of modeling, solving and optimization decision problems using the novel hybrid approach. Having combined the strengths of MILP and CP/CLP (Chaps. 2 and 3), we developed the environment that ensures the better and easier way of problem modeling and implementation and that provides the more effective search solution (Chap. 4). In order to verify the proposed approach, two illustrative examples are presented (Chap. 5).

## 2 Motivation

Based on [1–5], and our previous work [7, 9–13], we observed some advantages and disadvantages of these environments.

An integrated approach of constraint programming (CP) and mixed integer programming (MIP) can help to solve optimization problems that are intractable with either of the two methods alone [14–17]. Although operations research (OR) and constraint programming (CP) have different roots, the links between the two environments have grown stronger in recent years.

Both MIP/MILP/IP and finite domain CP/CLP involve variables and constraints. However, the types of the variables and constraints that are used, and the way the constraints are solved, are different in the two approaches [17].

MILP relies completely on linear equations and inequalities in integer variables, i.e., there are only two types of constraints: linear arithmetic (linear equations or inequalities) and integrity (stating that the variables have to take their values in the integer numbers). In finite domain CP/CLP, the constraint language is richer. In addition to linear equations and inequalities, there are various other constraints: disequalities, nonlinear, symbolic (*alldifferent, disjunctive, cumulative* etc.).

The motivation behind this work was to create a hybrid approach for supply chain modeling and optimization instead of using integer programming or constraint programming separately. We developed the hybrid framework for modeling and optimization of supply chain problems. In both MILP/MIP and CP/CLP, there is a group of constraints that can be solved with ease and a group of constraints that are difficult to solve. The easily solved constraints in MILP/MIP are linear equations and inequalities over rational numbers.

Integrity constraints are difficult to solve using mathematical programming methods and often the real problems of MIP/MILP make them NP-hard.

In CP/CLP, domain constraints with integers and equations between two variables are easy to solve. The system of such constraints can be solved over integer variables in polynomial time. The inequalities between two variables, general linear constraints (more than two variables), and symbolic constraints are difficult to solve, which makes real problems in CP/CLP NP-hard. This type of constraints reduces the strength of constraint propagation. As a result, CP/CLP is incapable of finding even the first feasible solution.

It follows from the above that what is difficult to solve in one environment can be easy to solve in the other.

The motivation was to offer the most effective tools for model-specific constraints and solution efficiency.

## 3 State of the Art

As mentioned in Chap. 1, the vast majority of decision-making models for the problems of production, logistics, supply chain are formulated in the form of mathematical programming (MIP, MILP, IP).

Due to the structure of these models (summing of discrete decision variables in the constraints and the objective function) and a large number of discrete decision variables (integer and binary) they can only be applied to small problems. Another disadvantage is that only linear constraints can be used. In practice, the issues related to the production, distribution and supply chain constraints are often logical, nonlinear, etc. For these reasons the problem was formulated in a new way.

In our approach to modeling and optimization of constrained search problems we proposed the optimization environment, where:

- knowledge related to the problem can be expressed as linear and logical constraints (implementing all types of constraints of the previous MILP/MIP models [11–13] and introducing new types of constraints (logical, nonlinear, symbolic etc.));
- the optimization model solved using the proposed framework can be formulated as a pure model of MILP/MIP or of CP/ CLP, or it can also be a hybrid model;
- the problem is modeled in CP/CLP, which is far more flexible than MIP/MILP/IP;
- the novel method of constraint propagation is introduced (obtained by transforming the optimization model to explore its structure);
- constrained domains of decision variables, new constraints and values for some variables are transferred from CP/CLP into MILP/MIP;
- the efficiency of finding solutions to the problems of larger sizes is increased.

As a result, we obtained the more effective search solution for a certain class of decision and optimization constrained problems.

# 4 Hybrid Environment

Both environments have advantages and disadvantages. Environments based on the constraints such as CLPs are declarative and ensure a very simple modeling of decision problems, even those with poor structures if any. The problem is described by a set of logical predicates. The constraints can be of different types (linear, non-linear, logical, binary, symbolic etc.). The CLP does not require any search algorithms. This feature is characteristic of all declarative backgrounds, in which modeling of the problem is also a solution, just as it is in Prolog, SQL, etc. The CLP seems perfect for modeling any decision problem.

In OR numerous models of decision-making have been developed and tested, particularly in the area of decision optimization. Constantly improved methods and mathematical programming algorithms, such as the simplex algorithm, branch and bound, branch-and-cost [21] etc., have become classics now.

The proposed method's strength lies in high efficiency of optimization algorithms and a substantial number of tested models. The decision problems we deal with in this paper, very common in manufacturing, logistics, supply chain, etc., have a number of decision variables, including binary and integer ones, which are aggregated in the constraints.

Traditional methods when used alone to solve complex problems provide unsatisfactory results. This is related directly to different treatment of variables and constraints in those approaches (Chap. 2). The proposed hybrid approach, a composition of methods as described in Chap. 3 offers the optimal system for specific contexts

## 4.1 Architecture and Implementation of Hybrid Environment (HE)

This Hybrid Environment (HE) consists of MIP/MILP/CLP/Hybrid models and solution framework to solve them. The concept of this environment with its phases (P1…P5, G1…G3) is presented in Fig. 1. A detailed description of the phases in the order of execution is shown in Table 1.

From a variety of tools for the implementation of the CP/CLP in HE, ECLiPSe software [22] was selected. ECLiPSe is an open-source software system for the costeffective development and deployment of constraint programming applications. Environment for the implementation of MILP/MIP/IP in HE was LINGO by LINDO Systems. LINGO Optimization Modeling Software is a powerful tool for building and solving mathematical optimization models [23].

ECLiPSe software is the environmental leader in HE. ECLiPSe was used to implement the following phases of the framework: P1, P2, P3, G1, G2, G3 (Fig. 1, Table 1.). The transformed files of the model were transferred from ECLiPSe to LINGO where they were merged (P4). Then the complete model was solved using LINGO efficient solvers (P5). Constraint propagation (phase-P3) greatly affected the efficiency of the solution. Therefore phase P2 was introduced. During this phase, the transformation

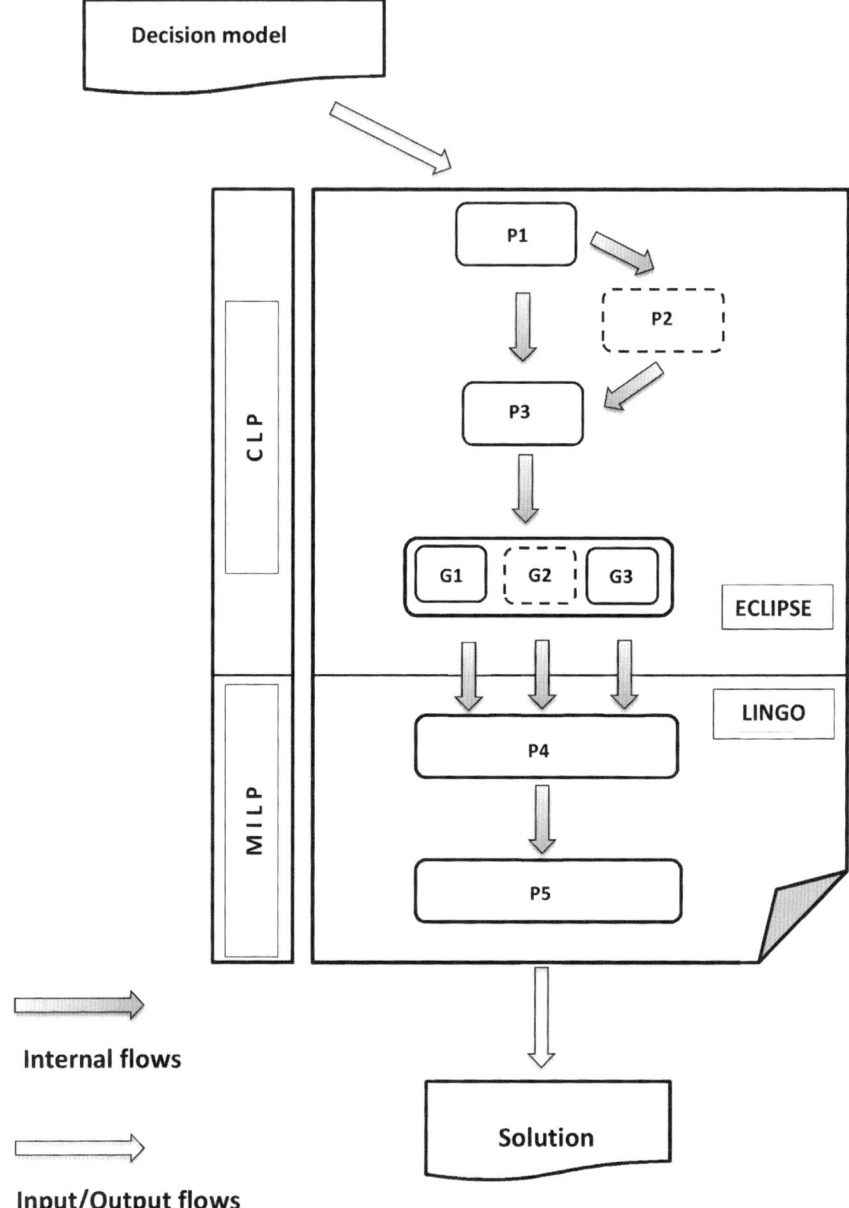

**Fig. 1** Detailed scheme of the Hybrid Environment (HE), optional phase marked by a dashed line

was performed using the structure and properties of the model. This is an optional phase that depends on the modeled problem. The details of this phase will be presented in one of the illustrative examples in Sect. 5 (cost optimization of supply chain).

**Table 1** Description of phases

| Phase | P1 |
|---|---|
| Name | Implementation of decision model |
| description | The implementation of the model in CLP, the term representation of the problem in the form of predicates |
| Phase | P2 |
| Name (optional) | Transformation of implemented model for better constraint propagation |
| description | The transformation of the original problem aimed at extending the scope of constraint propagation. The transformation uses the structure of the problem. The most common effect is a change in the representation of the problem by reducing the number of decision variables, and the introduction of additional constraints and variables, changing the nature of the variables, etc |
| Phase | P3 |
| Name | Constraint propagation |
| description | Constraint propagation for the model. Constraint propagation is one of the basic methods of CLP. As a result, the variable domains are narrowed, and in some cases, the values of variables are set, or even the solution can be found |
| Phase | G1 |
| Name | Generation of MILP/MIP/IP model |
| description | Generation of the model for mathematical programming. Generation performed automatically using CLP predicate. The resulting model is in a format accepted by the system LINGO |
| Phase | G2 |
| Name | Generation of additional constraints (optional) |
| description | Generation of additional constraints on the basis of the results obtained in step P3 |
| Phase | G3 |
| Name | Generation domains of decision variables and other values |
| description | Generation of domains for different decision variables and other parameters based on the propagation of constraints. Transmission of this information in the form of fixed value of certain variables and/or additional constraints to the MP |
| Phase | P4 |
| Name | Merging MILP/MIP/IP model |
| description | Merging files generated during the phases G1, G2, G3 into one file. It is a model file format in LINGO system |
| Phase | P5 |
| Name | Solving MILP/MIP/IP model |
| description | The solution model from the previous stage by LINGO. Generation of the report with the results and parameters of the solution |

# 5 Illustrative Examples

The proposed HE environment was verified and tested for two illustrative examples. The first example is the authors' original model of cost optimization of supply chain with multimodal transport (Sect. 5.1). The second is a 2E-CVRP model (Sect. 5.2). It is the known benchmark of a very large number of sets/instances of data and their solutions.

## 5.1 Cost Optimization of Supply Chain with Multimodal Transport

A detailed description of the cost optimization of supply chain models, their constraints, parameters and decision variables etc. are presented in [18] and Table 2.

During the first stage, the model was formulated as a MILP problem [10, 11], [18] in order to test the proposed environment (Fig. 1) against the classical integer-programming environment [23]. The next step involved the implementation and solving of the hybrid model. Indices, parameters and decision variables in the models together with their descriptions are provided in Table 2. The simplified structure of the supply chain network for this model, composed of producers, distributors and customers is presented in Fig. 2.

The proposed models are the cost models that take into account three other types of parameters, i.e., the spatial parameters (area/volume occupied by the product, distributor capacity and capacity of transport unit), time (duration of delivery and service by distributor, etc.) and the transport mode.

The main assumptions made for the construction of these models were as follows:

- the shared information process in the supply chain consists of resources (capacity, versatility, costs), inventory (capacity, versatility, costs, time), production (capacity, versatility, costs), product (volume), transport (cost, mode, time), demand, etc.;
- a part of the supply chain has the structure as in Fig. 2.;
- the transport is multimodal (several modes of transport, a limited number of means of transport for each mode);
- the environmental aspects of use of transport modes are taken into account;
- different products are combined in one batch of transport;
- the cost of supplies is presented in the form of a function (in this approach, linear function of fixed and variable costs);
- models have linear or linear and logical (hybrid model) constraints;
- logical constraints of hybrid model allow the distribution of exclusively one of two selected products in the distribution center and allow the production of exclusively one of two selected products in the factory.

**Table 2** Summary indices, parameters and decision variabless

| Symbol | Description |
| --- | --- |
| Indices | |
| $k$ | product type ($k = 1 \ldots O$) |
| $j$ | delivery point/customer/city ($j = 1 \ldots M$) |
| $i$ | manufacturer/factory ($i = 1 \ldots N$) |
| $s$ | distributor /distribution center ($s = 1 \ldots E$) |
| $d$ | mode of transport ($d = 1 \ldots L$) |
| $N$ | number of manufacturers/factories |
| $M$ | number of delivery points/customers |
| $E$ | number of distributors |
| $O$ | number of product types |
| $L$ | number of mode of transport |
| Input parameters | |
| $F_s$ | the fixed cost of distributor/distribution center $s$ |
| $P_k$ | the area/volume occupied by product $k$ |
| $V_s$ | distributor $s$ maximum capacity/volume |
| $W_{i,k}$ | production capacity at factory $i$ for product $k$ |
| $C_{i,k}$ | the cost of product $k$ at factory $i$ |
| $R_{s,k}$ | if distributors can deliver product k then $R_{sk} = 1$, otherwise $R_{sk} = 0$ |
| $Tp_{s,k}$ | the time needed for distributor $s$ to prepare the shipment of product $k$ |
| $Tc_{j,k}$ | the cut-off time of delivery to the delivery point/customer $j$ of product $k$ |
| $Z_{j,k}$ | customer demand/order $j$ for product $k$ |
| $Z_{td}$ | the number of transport units using mode of transport $d$ |
| $Pt_d$ | the capacity of transport unit using mode of transport $d$ |
| $Tf_{i,s,d}$ | the time of delivery from manufacturer i to distributor $s$ using mode of transport $d$ |
| $K1_{i,s,k,d}$ | the variable cost of delivery of product $k$ from manufacturer $i$ to distributor $s$ using mode of transport $d$ |
| $R1_{i,s,d}$ | if manufacturer $i$ can deliver to distributor $s$ using mode of transport $d$ then $R1_{i,s,d} = 1$, otherwise $R1_{i,s,d} = 0$ |
| $A_{i,s,d}$ | the fixed cost of delivery from manufacturer $i$ to distributor $s$ using mode of transport $d$ |
| $Koa_{i,s,d}$ | the total cost of delivery from manufacturer $i$ to distributor $s$ using mode of transport $d$ |
| $Tm_{s,j,d}$ | the time of delivery from distributor $s$ to customer $j$ using mode of transport $d$ |
| $K2_{s,j,k,d}$ | the variable cost of delivery of product $k$ from distributor $s$ to customer $j$ using mode of transport $d$ |
| $R2_{s,j,d}$ | if distributor $s$ can deliver to customer $j$ using mode of transport $d$ then $R2_{s,j,d} = 1$, otherwise $R2_{s,j,d} = 0$ |

(countinued)

**Table 2** (countinued)

| | |
|---|---|
| $Gs_{,j,d}$ | the fixed cost of delivery from distributor $s$ to customer $j$ using mode of transport $d$ |
| $Kogs_{,j,d}$ | the total cost of delivery from distributor $s$ to customer $j$ using mode of transport $d$ |
| $Odd$ | the environmental cost of using mode of transport $d$ |
| Decision variables | |
| $X_{i,s,k,d}$ | delivery quantity of product $k$ from manufacturer $i$ to distributor $s$ using mode of transport $d$ |
| $Xa_{i,s,d}$ | if delivery is from manufacturer $i$ to distributor $s$ using mode of transport $d$ then $Xa_{i,s,d} = 1$, otherwise $Xa_{i,s,d} = 0$ |
| $Xb_{i,s,d}$ | the number of courses from manufacturer $i$ to distributor $s$ using mode of transport $d$ |
| $Y_{s,j,k,d}$ | delivery quantity of product $k$ from distributor $s$ to customer $j$ using mode of transport $d$ |
| $Ya_{s,j,d}$ | if delivery is from distributor $s$ to customer $j$ using mode of transport $d$ then $Ya_{s,j,d} = 1$, otherwise $Ya_{s,j,d} = 0$ |
| $Yb_{s,j,d}$ | the number of courses from distributor $s$ to customer $j$ using mode of transport $d$ |
| $Tc_s$ | if distributor $s$ participates in deliveries, then $Tc_s = 1$, otherwise $Tc_s = 0$ |
| $CW$ | Arbitrarily large constant |

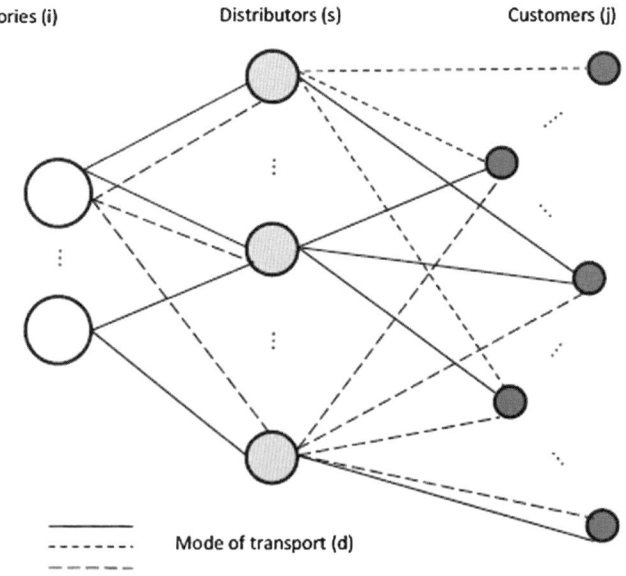

**Fig. 2** The simplified structure of the supply chain network

**Objective Function.** The objective function (1) defines the aggregate costs of the entire chain and consists of five elements. The first element comprises the fixed costs associated with the operation of the distributor involved in the delivery (e.g. distribution centre, warehouse, etc.). The second element corresponds to environmental costs of using various means of transport. Those costs are dependent on the number of courses of the given means of transport, and on the other hand, on the environmental levy, which in turn may depend on the use of fossil fuels and carbon-dioxide emissions.

The third component determines the cost of the delivery from the manufacturer to the distributor. Another component is responsible for the costs of the delivery from the distributor to the end user (the store, the individual client, etc.). The last component of the objective function determines the cost of manufacturing the product by the given manufacturer.

Formulating the objective function in this manner allows comprehensive cost optimization of various aspects of supply chain management. Each subset of the objective function with the same constrains provides a subset of the optimization area and makes it much easier to search for a solution.

$$\sum_{s=1}^{E} F_s T c_s + \sum_{d=1}^{L} Od_d (\sum_{i=1}^{N} \sum_{s=1}^{E} Xb_{i,s,d} + \sum_{s=1}^{E} \sum_{j=1}^{M} Yb_{j,s,d})$$

$$+ \sum_{i=1}^{N} \sum_{s=1}^{E} \sum_{d=1}^{L} Koa_{i,s,d} + \sum_{s=1}^{E} \sum_{j=1}^{M} \sum_{d=1}^{L} Kog_{s,j,d} + \sum_{i=1}^{N} \sum_{k=1}^{O} (C_{i,k} \sum_{s=1}^{E} \sum_{d=1}^{L} X_{i,s,k,d})$$

$$(1)$$

### 5.1.1 Constraints

The model was based on constraints (2)…(24) Constraint (2) specifies that all deliveries of product $k$ produced by the manufacturer $i$ and delivered to all distributors $s$ using mode of transport $d$ do not exceed the manufacturer's production capacity.

Constraint (3) covers all customer $j$ demands for product $k(Z_{j,k})$ through the implementation of delivery by distributors $s$ (the values of decision variables $Y_{i,s,k,d}$). The flow balance of each distributor $s$ corresponds to constraint (4). The possibility of delivery is dependent on the distributor's technical capabilities—constraint (5). Time constraint (6) ensures the terms of delivery are met. Constraints (7a), (7b), (8) guarantee deliveries with available transport taken into account. Constraints (9), (10), (11) set values of decision variables based on binary variables $T_{cs}$, $Xa_{i,s,d}$, $Ya_{s,j,d}$. Dependencies (12) and (13) represent the relationship based on which total costs are calculated. In general, these may be any linear functions. The remaining constraints (14)…(23) arise from the nature of the model (MILP).

Constraint (24) allows the distribution of exclusively one of the two selected products in the distribution center $s$. Similarly, constraint (25) allows the production of exclusively one of the two selected products in the factory $i$.

Those constraints result from technological, marketing, sales or safety reasons. Therefore, some products cannot be distributed and/or produced together. The constraint can be reused for different pairs of product $k$ and for some of or all distribution centers $s$ and factories $i$. A logical constraint like this cannot be easily implemented in a linear model. Only declarative application environments based on constraint satisfaction problem (CSP) make it possible to implement constraints such as (24), (25).

The addition of constraints of that type changes the model class. It is a hybrid model.

$$\sum_{s=1}^{E}\sum_{d=1}^{L} X_{i,s,k,d} \leq W_{i,k} \text{ for } i = 1\dots N, \qquad k = 1\dots 0 \tag{2}$$

$$\sum_{s=1}^{E}\sum_{d=1}^{L} Y_{s,j,k,d} \geq Z_{j,k} \text{ for } j = 1\dots M, \qquad k = 1\dots 0 \tag{3}$$

$$\sum_{i=1}^{N}\sum_{d=1}^{L} X_{i,s,k,d} = \sum_{k=1}^{M}\sum_{d=1}^{L} Y_{s,j,k,d} \text{ for } s = 1\dots E, \qquad k = 1\dots 0 \tag{4}$$

$$\sum_{k=1}^{O}(P_k(\sum_{i=1}^{N}\sum_{d=1}^{L} X_{i,s,k,d}) \leq Tc_s V_s \text{ for } s = 1\dots E \tag{5}$$

$$Xa_{i,s,d}Tf_{i,s,d} + Xa_{i,s,d}Tp_{s,k} + Ya_{s,j,d}Tm_{s,j,d} \leq Tc_{j,k} \tag{6}$$
$$\text{for } i = 1\dots N, \quad s = 1\dots E, \quad j = 1\dots M, \quad k = 1\dots O, \quad d = 1\dots L$$

$$R1_{i,s,d}Xb_{i,s,d}Pt_d \geq X_{i,s,k,d} \text{ for } i = 1\dots N, \quad s = 1\dots E, \quad k = 1\dots O, \quad d = 1\dots L \tag{7a}$$

$$R2_{s,j,d}Yb_{s,j,d}Pt_d \geq Y_{s,j,k,d} \text{ for } s = 1\dots E, \quad j = 1\dots M, \quad k = 1\dots O, \quad d = 1\dots L \tag{7b}$$

$$\sum_{i=1}^{N}\sum_{s=1}^{E} Xb_{i,s,d} + \sum_{j=1}^{M} Yb_{j,s,d} \leq Zt_d \text{ for } d = 1\dots L \tag{8}$$

$$\sum_{i=1}^{N}\sum_{d=1}^{L} \leq CwTc_s \text{ for } s = 1\dots E \tag{9}$$

$$Xb_{i,s,d} \leq CwXa_{i,s,d} \text{ for } i = 1\dots N, \quad s = 1\dots E, \quad d = 1\dots L \tag{10}$$

$$Yb_{s,j,d} \leq CwYa_{s,j,d} \text{ for } s = 1 \ldots E, \quad j = 1 \ldots M, \quad d = 1 \ldots L \quad (11)$$

$$Koa_{i,s,d} = A_{i,s,d} Xb_{i,s,d} + \sum_{k=1}^{O} K1_{i,s,k,d} X_{i,s,k,d} \quad (12)$$
$$\text{for } i = 1 \ldots N, \quad s = 1 \ldots E, \quad d = 1 \ldots L$$

$$Kog_{s,j,d} = G_{s,j,d} Yb_{s,j,d} + \sum_{k=1}^{O} K2_{s,j,k,d} Y_{s,j,k,d}$$
$$\text{for } s = 1 \ldots E, \quad j = 1 \ldots M, \quad d = 1 \ldots L \quad (13)$$

$$X_{i,s,k,d} \geq 0 \text{ for } i = 1 \ldots N, \quad s = 1 \ldots E, \quad k = 1 \ldots O, \quad d = 1 \ldots L \quad (14)$$

$$Xb_{i,s,d} \geq 0 \text{ for } i = 1 \ldots N, \quad s = 1 \ldots E, \quad d = 1 \ldots L \quad (15)$$

$$Yb_{s,j,d} \geq 0 \text{ for } s = 1 \ldots E, \quad j = 1 \ldots M, \quad d = 1 \ldots L \quad (16)$$

$$X_{i,s,k,d} \in C \text{ for } i = 1 \ldots N, \quad s = 1 \ldots E, \quad k = 1 \ldots O, \quad d = 1 \ldots L$$
$$(17)$$
$$Xb_{i,s,d} \in C \text{ for } i = 1 \ldots N, \quad s = 1 \ldots E, \quad d = 1 \ldots L \quad (18)$$

$$Y_{s,j,k,d} \in C \text{ for } s = 1 \ldots E, \quad j = 1 \ldots M, \quad k = 1 \ldots O, \quad d = 1 \ldots L$$
$$(19)$$
$$Yb_{s,j,d} \in 0 \text{ for } s = 1 \ldots E, \quad j = 1 \ldots M, \quad d = 1 \ldots L \quad (20)$$

$$Xa_{i,s,d} \in \{0, 1\} \text{ for } i = 1 \ldots N, \quad s = 1 \ldots E, \quad d = 1 \ldots L \quad (21)$$

$$Ya_{s,j,d} \in \{0, 1\} \text{ for } s = 1 \ldots E, \quad j = 1 \ldots M, \quad d = 1 \ldots L \quad (22)$$

$$Tc_s \in \{0, 1\} \text{ for } s = 1 \ldots E \quad (23)$$

$$Exclusion D(X_{i,s,k,d}, X_{i,s,l,d}, s) \text{ for } k \neq l, \quad s = 1 \ldots E \quad (24)$$

$$Exclusion P(X_{i,s,k,d}, X_{i,s,l,d}, i) \text{ for } k \neq l, \quad i = 1 \ldots N \quad (25)$$

**Model transformation.** Due to the nature of the decision problem (adding up decision variables and constraints involving a lot of variables), the constraint propagation efficiency decreases dramatically. Constraint propagation is one of the most important methods in CLP affecting the efficiency and effectiveness of the CLP and hybrid optimization environment (Fig. 1, Table 1). For that reason, research into more efficient and more effective methods of constraint propagation was conducted.

**Fig. 3** Representation of the
problem. **a** The classical
approach-definition. **b** The
classical approach—the
process of finding a solution

(**a**)
[O_n,P,M,D,F,Tu,Tu,Oq,X,T]

(**b**)
[[o_1,p1,m1,_,_,_,_,10,_,8],
[o_2,p1,m2,_,_,_,_,20,_,6],...]

The results included different representation of the problem and the manner of its implementation.

The classical problem modeling in the CLP environment consists in building a set of predicates with parameters.

Each CLP predicate has a corresponding multi-dimensional vector representation. While modeling both problems, quantities $i, s, k, d$ and decision variable $X_{i,s,k,d}$ were vector parameters (Fig. 3a). As shown in Fig. 3b, for each vector there were 5 values to be determined, defining the size of the delivery, factories, distributors involved in the delivery and the mode of transport.

The process of finding the solution may consist in using the constraint propagation methods, variable labeling and the backtracking mechanism. The numbers of parameters that must be specified/labeled in the given predicate/vector critically affect the quality of constraint propagation and the number of backtracks. In both models presented above, the classical problem representation included five parameters: $i, s, k, d$ and $X_{i,s,k,d}$. Considering the domain size of each parameter, the process was complex and time-consuming. In addition, the above representation (Figs. 3a and 3b) arising from the structure of the problem is the cause of many backtracks.

Our idea involved the transformation of the problem by changing its representation without changing the very problem. All permissible routes were first generated based on the fixed data and a set of orders, then the specific values of parameters $i, s, k, d$ were assigned to each of the routes. Other parameters of the new representation of the problem relate to the delivery time, fixed and variable costs and the maximum demand for the product are determined by the CLP predicates based on fixed data. In this way, only decision variables $X_{i,s,k,d}$ (deliveries) had to be specified (Fig. 4). This transformation fundamentally improved the efficiency of the constraint propagation and reduced the number of backtracks. A route model is a name adopted for the models that underwent the transformation.

Symbols necessary to understand both the representation of the problem and their descriptions are presented in Table 3

**Fig. 4** Representation of the
problem in the novel
approach-set of feasible
routes

[[route_1,f1,p1,c1,m1,s1,s1,5,12,100,_],
[route_2,f1,p1,c1,m1,s1,s2,6,14,100,_],
[route_3,f1,p1,c1,m1,s2,s1,6,22,100,_], ...]

**Table 3** Symbols used in the representation of the problem

| Symbol | Description |
|--------|-------------|
| O_n | order number |
| P | products, $P \in \{p_1, p_2, ..., p_o\}$ |
| M | customers, $M \in \{m_1, m_2, ..., m_m\}$ |
| D | distributors, $D \in \{c_1, c_2, ..., c_e\}$ |
| F | factories, $F \in \{f_1, f_2, ..., f_n\}$ |
| Tu | transport unit, $Tu \in \{s_1, s_2, ..., s_l\}$ |
| T | delivery time/period |
| Oq | order quantity |
| X | delivery quantity |
| route_n | routes name-number |

## 5.2 Two-Echelon Capacitated Vehicle Routing Problem

The 2E-CVRP is proposed as a benchmark verifying the presented approach. The Two-Echelon Capacitated Vehicle Routing Problem (2E-CVRP) is an extension of the classical Capacitated Vehicle Routing Problem (CVRP) where the delivery depot-customers pass through intermediate depots (called satellites). As in CVRP, the goal is to deliver goods to customers with known demands, minimizing the total delivery cost in the respect of vehicle capacity constraints. Multi-echelon systems presented in the literature usually explicitly consider the routing problem at the last level of the transportation system, while a simplified routing problem is considered at higher levels [19, 20, 24].

In 2E-CVRP, the freight delivery from the depot to the customers is managed by shipping the freight through intermediate depots. Thus, the transportation network is decomposed into two levels (Fig. 5): the 1st level connecting the depot (d) to intermediate depots (s) and the 2nd one connecting the intermediate depots (s) to the customers (c). The objective is to minimize the total transportation cost of the vehicles involved in both levels. Constraints on the maximum capacity of the vehicles and the intermediate depots are considered, while the timing of the deliveries is ignored.

From a practical point of view, a 2E-CVRP system operates as follows (Fig. 5):

- freight arrives at an external zone, the depot, where it is consolidated into the 1st-level vehicles, unless it is already carried into a fully-loaded 1st-level vehicles;
- ach 1st-level vehicle travels to a subset of satellites that will be determined by the model and then it will return to the depot;
- at a satellite, freight is transferred from 1st-level vehicles to 2nd-level vehicles.

The mathematical model (MILP) was taken from [18]. It required some adjustments and error corrections. Table 4 shows the parameters and decision variables of 2E-CVRP. Figure 5 shows an example of the 2E-CVRP - transportation network. The transformation of this model in the hybrid approach focused on the resizing of Yk,i,j decision variable by introducing additional imaginary volume of freight shipped from the satellite and re-delivered to it. Such transformation resulted in two facts.

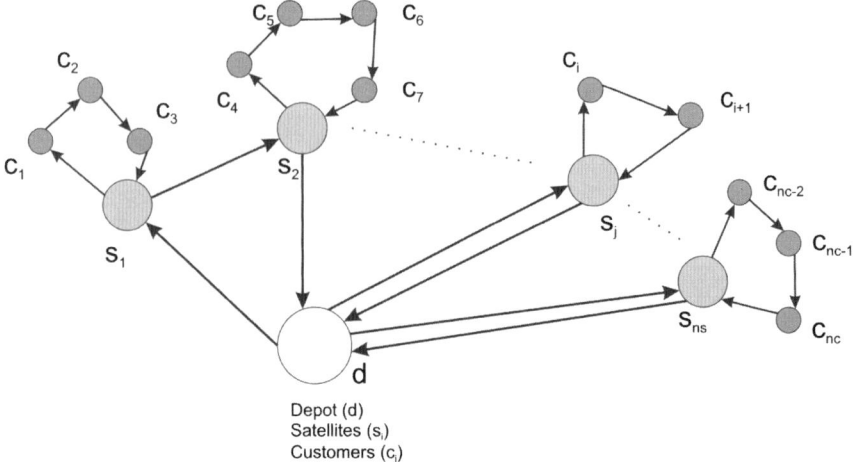

**Fig. 5**  Example of 2E-CVRP transportation network

First of all, it forced the vehicle to return to the satellite from which it started its
trip. Secondly, it reduced decision variable $Y_{k,i,j}$ to variable $Y_{i,j}$ which decreased
the size of the combinatorial problem.

## 6  Numerical Experiments

The proposed HE environment was verified and tested for two illustrative examples.
The first example is the authors' original model of cost optimization of supply chain
with multimodal transport (Chap. 5.1). The second is a 2E-CVRP model (Chap. 5.2).
It is the known benchmark of a very large number of sets/instances of data and their
solutions.

### 6.1  Cost Optimization of Supply Chain
###      with Multimodal Transport

In order to verify and evaluate the proposed approach, many numerical experiments
were performed. All the examples relate to the supply chain with seven manufacturers
$(i = 1 \ldots 7)$, three distributors $(s = 1 \ldots 3)$, ten customers $(j = 1 \ldots 10)$, three
modes of transport $(d = 1 \ldots 3)$, and twenty types of products $(k = 1 \ldots 20)$.

The first series of experiments was designed to show the advantages of the hybrid
approach used.

**Table 4** Summary indices, parameters and decision variables

| Symbol | Description |
|---|---|
| **Indices** | |
| $n_s$ | Number of satelites |
| $n_c$ | Number of customers |
| $V_0 = \{v_o\}$ | Deport |
| $V_s = \{vs_1, ..., vs_{ns}\}$ | Set of satellites |
| $V_c = \{vc_1, ..., vc_{nc}\}$ | Set of customers |
| **Input parameters** | |
| $m_1$ | Number of the 1st-level satelites |
| $M_2$ | Number of the 2nd-level satelites |
| $k_1$ | Capacity of the vehicles for the 1st level |
| $k_2$ | Capacity of the vehicles for the 2nd level |
| $d_i$ | Demand required by customer $i$ |
| $c_{i,j}$ | Cost of the arc $(i, j)$ |
| $s_k$ | Cost of loading/unloading operations of a unit of freight in satelite $k$ |
| **Decision variables** | |
| $X_{i,j}$ | Is an integer variable of the 1st-level routing and is equal to the number of 1st-level vehicles using arc $(i, j)$. |
| $Y_{k,i,j}$ | Is a binary variable of the 2nd-level routing and is equal to 1 if a 2ndlevel vehicle makes a route starting from satellite $k$ and goes from node $i$ to node $j$ and 0 otherwise |
| $Q1_{i,j}$ | Freight flow arc $(i, j)$ for the 1st-level |
| $Q2_{k,i,j}$ | Freight arc $(i, j)$ where $k$ represents the satellite where the freight is passing through. |
| $z_{k,j}$ | Binary variable that is equal to 1 if the freight to be delivered to customer $j$ is consolidated in satellite $k$ and 0 otherwise |

The experiments began with six examples: E1…E6 for the problem formulated in MILP (Chap. 5) [18]. Two approaches were used to implement the proposed model: mathematical programming (LINGO) and the hybrid approach (LINGO, ECLiPSe, transformation). The examples E1…E6 varied in terms of the number of orders (No). The set of all orders for calculation examples are given in Appendix. The experiments were conducted to optimize examples E7, E8, which are implementations of the hybrid model (with logical constraints) in the hybrid approach. The implementation of logic constraints for the hybrid model was as follows: product k = 1 cannot be distributed with product k = 7; product k = 2 cannot be distributed with product k = 18, and these products cannot be produced together. The results in the form of the objective function, the computation time, the number of discrete decision variables and constraints are shown in Table 5.

The analysis of the outcome indicates that the hybrid approach provided better results in terms of the time needed to find the solution in each case, and to obtain the

**Table 5** The results of numerical examples for both approaches

| E(No) | MILP-LINGO | | | | MILP-Hybrid | | | |
|---|---|---|---|---|---|---|---|---|
| | Fc | T | V(int) | C | Fc | T | V(int) | C |
| E1(5) | 717 | 23 | 6881(6429) | 5326 | 717 | 2 | 221 (204) | 537 |
| E2(10) | 2290* | 600** | 6881(6429) | 6271 | 2287 | 3 | 384 (354) | 550 |
| E3(20) | 3643* | 600** | 6881(6429) | 8161 | 3386 | 23 | 499 (454) | 565 |
| E4(40) | 6224* | 600** | 6881(6429) | 11941 | 5296 | 180 | 793 (746) | 566 |
| E5(80) | 11000* | 600** | 6881(6429) | 19501 | 9164 | 230 | 1313 (1267) | 566 |
| E6(100) | — | 600** | 6881(6429) | 23281 | 10915 | 256 | 1624 (1577) | 566 |
| P(No) | Hybrid-Hybrid | | | | | | | |
| | Fc | T | V(int) | C | | | | |
| E7(10) | 2805 | 4 | 386 (356) | 556 | | | | |
| E8(100) | 12575 | 384 | 1626 (1579) | 573 | | | | |
| Fc | the optimal value of the objective function | | | | | | | |
| T | solution finding time | | | | | | | |
| C | the number of constraints | | | | | | | |
| V(int) | the number of variables (integer variables) | | | | | | | |
| * | the feasible value of the objective function after the time T | | | | | | | |
| ** | calculation was stopped after 600s | | | | | | | |
| — | no feasible solution is found | | | | | | | |

optimal solution in some cases, which was impossible to do within the acceptable time limits using the traditional approaches.

## 6.2 Two-Echelon Capacitated Vehicle Routing Problem

For the final validation of the proposed hybrid approach, the benchmark (2E-CVRP) was selected. 2E-CVRP, a well described and widely discussed problem, corresponded to the issues to which our hybrid approach was applied.

The instances for computational examples were built from the existing instances for CVRP [25] denoted as E-n13-k4. All the instance sets can be downloaded from the website [26]. The instance set was composed of 5 small-sized instances with 1 depot, 12 customers and 2 satellites. The full instance consisted of 66 small-sized instances because the two satellites were placed over twelve customers in all 66 possible ways (number of combinations: 2 out of 12).

All the instances had the same position for depot and customers, whose coordinates were the same as those of instance E-n13-k4. Small-sized instances differed in the choice of two customers who were also satellites (En13-k4-2 (1,3), En13-k4-6 (1,6), En13-k4-61 (9,10) etc.).

The analysis of the results for the benchmark instances demonstrates that the hybrid approach may be a superior approach to the classical mathematical programming. For all examples, the solutions were found 2-16 times faster than they are in the classical approach.

As the presented benchmark was formulated as a MILP problem, the HE was tested for the solution efficiency. Owing to the hybrid approach the 2E-CVRP models can be extended over logical, nonlinear, and other constraints.

## 7 Conclusion and Discussion on Possible Extension

The efficiency of the proposed approach is based on the reduction of the combinatorial problem and using the best properties of both environments. The hybrid approach (Tables 5 and 6) makes it possible to find better solutions in the shorter time.

In addition to solving larger problems faster, the proposed approach provides virtually unlimited modeling options.

Therefore, the proposed solution is recommended for decision-making problems in the supply chains, logistics, transportation and manufacturing that have a similar structure to the presented models (Chap. 5). This structure is characterized by the constraints and objective function in which the many decision variables are added together. Further work will focus on running the optimization models with non-linear and logical constraints, multi-objective, uncertainty etc. in the hybrid optimization framework.

**Table 6** The results of numerical examples for both approaches

| E-n13-k4 | MILP-LINGO | | | | MILP-Hybrid | | | |
|---|---|---|---|---|---|---|---|---|
| | Fc | T | V | C | Fc | T | V | C |
| En13-k4-2 | 286 | 40371 | 368 | 1262 | 286 | 8720 | 186 | 1024 |
| En13-k4-6 | 230 | 125 | 368 | 1262 | 230 | 55 | 186 | 1024 |
| En13-k4-9 | 244 | 153 | 368 | 1262 | 244 | 44 | 186 | 1024 |
| En13-k4-20 | 276 | 535 | 368 | 1262 | 276 | 32 | 186 | 1024 |
| En13-k4-61 | 338 | 6648 | 368 | 1262 | 338 | 407 | 186 | 1024 |
| Fc | the optimal value of the objective function | | | | | | | |
| T | time of finding solution | | | | | | | |
| V/C | the number of integer variables/constraints | | | | | | | |
| * | the feasible value of the objective function after the time T | | | | | | | |

# Appendix

See Table 7

**Table 7** The results of numerical examples for both approaches

| Name | k | j | $T_{k,j}$ | $Z_{k,j}$ | Name | k | j | $T_{k,j}$ | $Z_{k,j}$ | Name | k | j | $T_{k,j}$ | $Z_{k,j}$ |
|------|-----|-----|-----|-----|------|-----|-----|-----|-----|------|-----|-----|-----|-----|
| z101 | p01 | m1 | 25 | 8  | z405 | p05 | m4 | 70 | 15 | z714 | p14 | m7 | 10 | 10 |
| z102 | p02 | m1 | 15 | 10 | z409 | p09 | m4 | 10 | 10 | z717 | p17 | m7 | 30 | 10 |
| z104 | p04 | m1 | 15 | 10 | z410 | p10 | m4 | 20 | 10 | z718 | p18 | m7 | 30 | 10 |
| z106 | p06 | m1 | 30 | 10 | z412 | p12 | m4 | 10 | 10 | z801 | p01 | m8 | 20 | 10 |
| z108 | p08 | m1 | 45 | 10 | z414 | p14 | m4 | 30 | 10 | z802 | p02 | m8 | 10 | 10 |
| z111 | p11 | m1 | 15 | 8  | z415 | p15 | m4 | 45 | 10 | z808 | p08 | m8 | 45 | 8  |
| z112 | p12 | m1 | 75 | 10 | z505 | p05 | m5 | 60 | 10 | z809 | p09 | m8 | 55 | 10 |
| z116 | p16 | m1 | 30 | 10 | z506 | p06 | m5 | 30 | 10 | z812 | p12 | m8 | 20 | 10 |
| z119 | p19 | m1 | 10 | 10 | z507 | p07 | m5 | 20 | 10 | z813 | p13 | m8 | 20 | 10 |
| z120 | p20 | m1 | 10 | 10 | z508 | p08 | m5 | 30 | 10 | z814 | p14 | m8 | 10 | 10 |
| z202 | p02 | m2 | 40 | 8  | z509 | p09 | m5 | 40 | 10 | z815 | p15 | m8 | 10 | 10 |
| z203 | p03 | m2 | 50 | 10 | z510 | p10 | m5 | 50 | 10 | z818 | p18 | m8 | 25 | 8  |
| z205 | p05 | m2 | 30 | 10 | z515 | p15 | m5 | 45 | 8  | z819 | p19 | m8 | 25 | 10 |
| z206 | p06 | m2 | 30 | 10 | z516 | p16 | m5 | 30 | 10 | z906 | p06 | m9 | 20 | 10 |
| z207 | p07 | m2 | 30 | 10 | z519 | p19 | m5 | 10 | 10 | z907 | p07 | m9 | 10 | 10 |
| z208 | p08 | m2 | 30 | 10 | z520 | p20 | m5 | 20 | 10 | z908 | p08 | m9 | 10 | 8  |
| z212 | p12 | m2 | 35 | 10 | z601 | p01 | m6 | 10 | 10 | z909 | p09 | m9 | 30 | 10 |
| z213 | p13 | m2 | 30 | 8  | z602 | p02 | m6 | 20 | 10 | z912 | p12 | m9 | 25 | 10 |
| z214 | p14 | m2 | 10 | 10 | z603 | p03 | m6 | 20 | 10 | z913 | p13 | m9 | 20 | 10 |
| z215 | p15 | m2 | 10 | 10 | z604 | p04 | m6 | 20 | 10 | z914 | p14 | m9 | 10 | 10 |
| z303 | p03 | m3 | 10 | 8  | z606 | p06 | m6 | 50 | 8  | z915 | p15 | m9 | 10 | 10 |
| z304 | p04 | m3 | 30 | 8  | z607 | p07 | m6 | 15 | 10 | z919 | p19 | m9 | 40 | 10 |
| z306 | p06 | m3 | 20 | 10 | z608 | p08 | m6 | 20 | 10 | z920 | p20 | m9 | 40 | 10 |
| z307 | p07 | m3 | 30 | 10 | z609 | p09 | m6 | 10 | 10 | zA01 | p01 | mA | 20 | 10 |
| z313 | p13 | m3 | 25 | 7  | z616 | p16 | m6 | 50 | 25 | zA02 | p02 | mA | 10 | 10 |
| z314 | p14 | m3 | 25 | 10 | z617 | p17 | m6 | 45 | 10 | zA10 | p10 | mA | 30 | 8  |
| z317 | p17 | m3 | 20 | 10 | z701 | p01 | m7 | 10 | 10 | zA11 | p11 | mA | 35 | 10 |
| z318 | p18 | m3 | 30 | 10 | z702 | p02 | m7 | 20 | 10 | zA14 | p14 | mA | 10 | 10 |
| z319 | p19 | m3 | 40 | 10 | z707 | p07 | m7 | 25 | 10 | zA15 | p15 | mA | 10 | 10 |
| z320 | p20 | m3 | 50 | 10 | z708 | p08 | m7 | 30 | 10 | zA16 | p16 | mA | 20 | 10 |
| z401 | p01 | m4 | 40 | 10 | z711 | p11 | m7 | 10 | 10 | zA17 | p17 | mA | 20 | 10 |
| z402 | p02 | m4 | 50 | 10 | z712 | p12 | m7 | 20 | 10 | zA19 | p19 | mA | 15 | 10 |
| z403 | p03 | m4 | 60 | 10 | z713 | p13 | m7 | 20 | 10 | zA20 | p20 | mA | 35 | 10 |
| z404 | p04 | m4 | 30 | 10 |      |     |     |    |    |      |     |     |    |    |

# References

1. Kanyalkar, A.P., Adil, G.K.: An integrated aggregate and detailed planning in a multi-site production environment using linear programming. Int. J. Prod. Res. **43**, 4431–4454 (2005)
2. Perea-lopez, E., Ydstie, B.E., Grossmann, I.E.: A model predictive control strategy for supply chain optimization. Comput. Chem. Eng. **27**, 1201–1218 (2003)
3. Christian Lang, J.: Production and Inventory Management with Substitutions, Production and Operations Management: Models and algorithms. Lecture Notes in Economics and Mathematical Systems. Springer, Berlin Heidelberg (2010)
4. Dang, Q., Nielsen, I., Steger-Jensen, K., Madsen, O.: Scheduling a single mobile robot for part-feeding tasks of production lines. J. Intell. Manuf. **20**(2), 211–221 (2013). doi:10.1007/s10845-013-0729-y
5. Mula, J., Peidro, D., Diaz-Madronero, M., Vicens, E.: Mathematical programming models for supply chain production and transport planning. Eur. J. Oper. Res. **204**, 377–390 (2010)
6. Apt, K., Wallace, M.: Constraint Logic Programming using Eclipse. Cambridge University Press, New York (2006)
7. Sitek, P., Wikarek, J.: A declarative framework for constrained search problems, new frontiers in applied artificial intelligence. In: Nguyen, N.T., et al. (eds.) New Frontiers in Applied Artificial Intelligence. Lecture Notes in Artificial Intelligence, pp. 728–737. Springer, Berlin (2008)
8. Bocewicz, G., Banaszak, Z.: Declarative approach to cyclic steady states space refinement: periodic processes scheduling. Int. J. Adv. Manuf. Technol. **67**(1–4), 137–155 (2013)
9. Sitek, P.: Grouping products in a follow-up production control system for parallel partitioned flow production lines. In: Intelligent Manufacturing Systems IMS 2001: 6th IFAC Workshop, pp. 122–126. Pergamon, New York (2001)
10. Sitek, P., Wikarek, J.: The concept of decision support system structures for the distribution center. MPER (Manage. Prod. Eng. Rev.) **1**(3), 63–69 (2010)
11. Sitek, P., Wikarek, J.: Cost Optimization of Supply Chain with Multimodal Transport, Federated Conference on Computer Science and Information Systems (FedCSIS), pp. 1111–1118. IEEE, Wroclaw (2012)
12. Sitek, P., Wikarek, J.: Supply chain optimization based on a MILP model from the perspective of a logistics provider. MPER (Manag. Prod. Eng. Rev.) **3**(2), 49–61 (2012)
13. Sitek, P., Wikarek, J.: The declarative framework approach to decision support for constrained search problems, INTECH. 163–182 (2011)
14. Jain, V., Grossmann, I.E.: Algorithms for hybrid MILP/CP models for a class of optimization problems. INFORMS J. Comput. **13**(4), 258–276 (2001)
15. Milano, M., Wallace, M.: Integrating operations research in constraint programming. Ann. Oper. Res. **175**(1), 37–76 (2010)
16. Achterberg, T., Berthold, T., Koch, T., Wolter, K.: Constraint integer programming: a new approach to integrate CP and MIP. In: Perron, L., Trick, M.A. (eds.) Integration of AI and OR Techniques in Constraint Programming for Combinatorial Optimization Problems. Lecture Notes in Computer Science, pp. 6–20. Springer, Berlin (2008)
17. Bockmayr, A., Kasper, T.: Branch-and-Infer, A Framework for Combining CP and IP. In: Milano, M. (ed.) Constraint and Integer Programming. Operations Research/Computer Science Interfaces Series, pp. 59–87. Springer, New York (2004)
18. Sitek, P., Wikarek, J.: A hybrid approach to supply chain modeling and optimization. In: Federated Conference on Computer Science and Information Systems (FedCSIS), pp. 1223–1230. Warsaw, Poland (2013)
19. Perboli, G., Tadei, R., Vigo, D.: The two-echelon capacitated vehicle routing problem: models and math-based heuristics. Transp. Sci. **45**, 364–380 (2011)
20. Crainic, T., Ricciardi, N., Storchi, G.: Advanced freight transportation systems for congested urban areas. Transp. Res. Part C **12**, 119–137 (2004)
21. Schrijver, A.: Theory of Linear and Integer Programming. Wiley, New York (1998). ISBN: 0-471-98232-6

22. Eclipse—The Eclipse foundation open source community website. http://www.eclipse.org (2014). Accessed 12 Aug 2014
23. Lindo Systems INC—LINDO™ Software for Integer Programming, Linear Programming, Nonlinear Programming, Stochastic Programming, Global Optimisation. http://www.lindo.com (2014). Accessed 12 Aug 2014
24. Ricciardi, N., Tadei, R., Grosso, A.: Optimal facility location with random throughput costs. Comput. Oper. Res. **29**(6), 593–607 (2002)
25. Christofides, N., Elion, S.: An algorithms for the vehicle dispatching problem. Oper. Res. Q. **20**, 309–318 (1969)
26. Orgroup—ORO Group Web-page. http://www.orgroup.polito.it (2014). Accessed 12 Aug 2014
27. Rossi, F., Van Beek, P., Walsh, T.: Handbook of Constraint Programming (Foundations of Artificial Intelligence). Elsevier Science Inc., New York (2006)

# Biased Random Key Genetic Algorithm for Multi-user Earth Observation Scheduling

**Panwadee Tangpattanakul, Nicolas Jozefowiez and Pierre Lopez**

**Abstract** This paper presents a biased random key genetic algorithm, or BRKGA, for solving a multi-user observation scheduling problem. BRKGA is an efficient method in the area of combinatorial optimization. It is usually applied to single objective problem. It needs to be adapted for multi-objective optimization. This paper considers two adaptations. The first one presents how to select the elite set, i.e., good solutions in the population. We borrow the elite selection methods from efficient multi-objective evolutionary algorithms. For the second adaptation, since the multi-objective optimization needs a set of solutions on the Pareto front, we investigate the idea to obtain several solutions from a single chromosome. Experiments are conducted on realistic instances, which concern the multi-user observation scheduling of an agile Earth observing satellite.

**Keywords** Multi-objective optimization · Scheduling · Earth observing satellite · Genetic algorithm

## 1 Introduction

This paper proposes two adaptations of the biased random key genetic algorithm for solving multi-objective optimization problems. They are elite selection methods

P. Tangpattanakul (✉)
Geo-Informatics and Space Technology Development Agency (GISTDA),
120 The Government Complex, Chaeng Wattana Road, Lak Si, Bangkok 10210, Thailand
e-mail: panwadee@gistda.or.th

N. Jozefowiez · P. Lopez
CNRS, LAAS, 7 avenue du Colonel Roche, 31400 Toulouse, France
e-mail: nicolas.jozefowiez@laas.fr

P. Lopez
e-mail: pierre.lopez@laas.fr

N. Jozefowiez
INSA, LAAS, Univ de Toulouse, 31400 Toulouse, France

P. Lopez
LAAS, Univ de Toulouse, 31400 Toulouse, France

S. Fidanova (ed.), *Recent Advances in Computational Optimization*,
Studies in Computational Intelligence 580, DOI 10.1007/978-3-319-12631-9_9

and a hybrid decoding. We experiment on instances of a multi-user observation scheduling problem for an agile Earth observing satellite (EOS).

The biased random key genetic algorithm (BRKGA) was first presented in [6]. The BRKGA combines the concept of random key and the principles of genetic algorithms. It operates on several individuals in a population. Each individual contains a chromosome, which is formed by several genes. The chromosome is encoded as a key vector of random real values in the interval [0, 1]. Generally, the random key vector represents one solution, which is obtained through the application of a decoder. It is an operator that builds a feasible solution for the problem from the chromosome. It is the only ad-hoc part of the method and must be deterministic.

In recent years, BRKGA was used to solve combinatorial optimization problems in various domains (e.g., communication, transportation, scheduling) [7]. For example, BRKGA was applied to solve the fiber installation in an optical network optimization problem [8]. The objective function was to minimize the cost of the optical components necessary to operate the network. In [11], a resource-constrained project scheduling problem with makespan minimization was solved by BRKGA. Nevertheless, all these works address single objective optimization problems. This paper considers multi-objective optimization. Several real world problems, e.g., in the area of engineering research and design, can be modeled as multi-objective optimization problems. When several objectives are considered simultaneously, a method to select the elite set has to be defined. It is the purpose of the first adaptation. For the second adaptation, we investigate an idea to improve the efficiency of BRKGA for solving multi-objective optimization problems. We propose the use of a decoder that does not return a single solution but several potentially-efficient solutions. These adaptations are experimented on the multi-user observation scheduling problem for an agile Earth observing satellite.

The mission of Earth observing satellites (EOSs) is to obtain photographs of the Earth surface satisfying user requirements. When the ground station center receives requests from users, it has to manage the requirements by selecting and scheduling a subset of photographs and transmit the schedule, which consists of a sequence of selected photographs, to the satellites. We consider an agile satellite, which has only one on-board camera that can move around three axes: roll, pitch, and yaw. The starting time of each photograph is not fixed; it can slide within a given visible time interval. The problem description of the agile EOS scheduling problem is presented in the ROADEF 2003 challenge [13]. This challenge required the scheduling solutions that maximize total profit of the acquired photographs for a single user and have to satisfy all physical constraints of agile EOSs. Algorithms based on simulated annealing [10] and Tabu search [4] were particularly proposed for this challenge.

Overall, we follow the description of the problem introduced in the ROADEF 2013 challenge. Two possible shapes of area can be required: spot or polygon. The polygon is a big area that the camera cannot take instantaneously. Hence it has to be decomposed into several strips of fixed width but variable length, as shown in Fig. 1. Among two possible acquisition directions, one acquisition can be selected for each strip. Two types of photograph can be required: a mono photograph is taken only

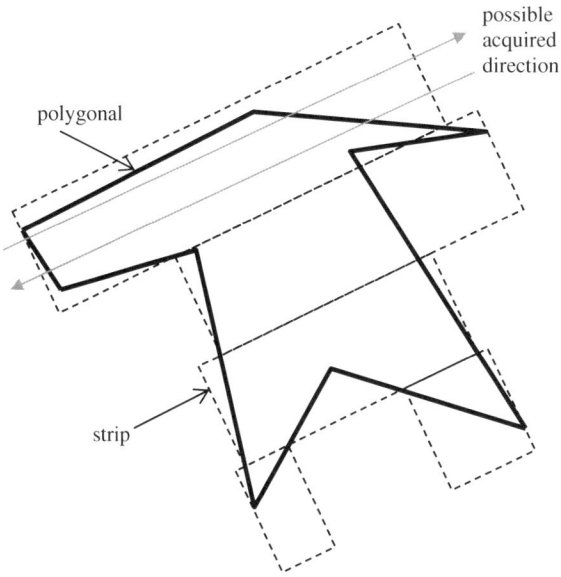

possible
acquired
direction

polygonal

strip

**Fig. 1** A polygon is decomposed into several strips; one of two possible directions can be selected for the acquisition of each strip

once, whereas a stereo photograph should be acquired twice in the same direction but from different angles.

For each acquisition, a time window is computed during which the photograph can be acquired. Moreover, a sufficient transition time must also be respected between two consecutive selected acquisitions.

The originality of our work lies in the consideration of multi-user requests leading to the definition of a bi-objective problem. The ground station center should maximize the total profit of the acquired photographs and simultaneously share fairly the satellite resources for all users by minimizing the maximum profit difference between users. For each solution, the two objective function values can be calculated by using a piecewise linear function of gain. This function is associated with a partial acquisition of the acquired request, as illustrated in Fig. 2. In [12], we proposed a straightforward use of biased random-key genetic algorithm (BRKGA). Note that multiple users have also been considered in [3]. However in this paper, the authors consider only a single objective.

The article is organized as follows. Section 2 explains how BRKGA is used to solve multi-objective optimization problems. The hybrid decoding allowing the association of one chromosome with several solutions is proposed in Sect. 3. Section 4 reports the computational results. Finally, conclusions and perspectives are discussed in Sect. 5.

**Fig. 2** Piecewise linear
function of gain $P(x)$
depending on the effective
ratio $x$ of acquired area [13]

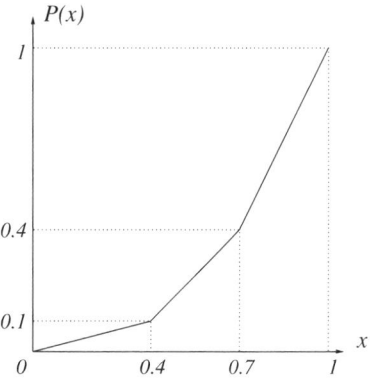

## 2 Biased Random Key Genetic Algorithm for Multi-objective Optimization Problems

A genetic algorithm (GA) is a metaheuristic method, which operates on several individuals in a population. Individuals should spread through the search space. The genetic algorithm uses the concept of survival of the fittest to converge toward optimal solutions. Each individual consists of a chromosome, which represents a solution. A GA starts by generating an initial population of size $p$. At each generation, selection, crossover, and mutation operators are applied. The generations are repeated until a stopping criterion is satisfied.

The biased random key genetic algorithm (BRKGA) was first presented in [6]. The BRKGA has different ways to select two parents for the crossover operation, compared with the original of random key genetic algorithm (RKGA) [1]. For BRKGA, the random key chromosome is formed by several genes, which are encoded by real values in the interval [0, 1]. Then, the chromosome is decoded in order to obtain a solution for the problem. The decoding strategy is problem dependent. The fitness value of a solution is computed in this decoding step. The current population is divided into two groups according to the fitness value. The best $p_e$ chromosomes from the current population are selected to become the elite set. The remaining chromosomes form the group of non-elite chromosomes.

The new population is composed of three different solution groups, as shown in Fig. 3. The first part is the elite set. The second part is a set of $p_m$ chromosomes, which are randomly generated to avoid the entrapment in a local optimum. These chromosomes are called mutant. The last part is filled by generating offspring using the crossover operator applied to one solution from the elite set and one solution from the non elite set. Each gene in the offspring is equal to the gene of the elite parent with the probability $\rho_e$. Otherwise, it is copied from the non-elite parent. Hence, the size of crossover offspring set is equal to $p - p_e - p_m$. The recommended parameter value setting is displayed in Table 1.

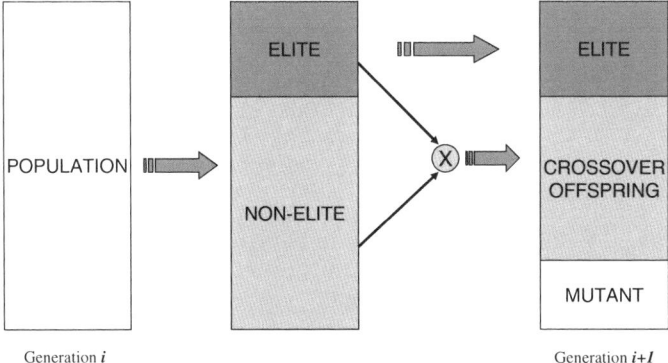

**Fig. 3** The population of the new generation by using BRKGA

**Table 1** Recommended parameter values of BRKGA [7]

| Parameter | Recommended value |
|---|---|
| $p$ | $p = a.n$, where $1 \leq a \in \mathbb{R}$ is a constant and $n$ is the length of the chromosome |
| $p_e$ | $0.10p \leq p_e \leq 0.25p$ |
| $p_m$ | $0.10p \leq p_m \leq 0.30p$ |
| $\rho_e$ | $0.5 \leq \rho_e \leq 0.8$ |

More details on the standard procedure for BRKGA can be found in [7]. BRKGA was applied to solve optimization problems arising in several applications. However, all problems consider only one objective. In this work, we study BRKGA for solving a multi-objective optimization problem. We will now explain how to adapt the standard BRKGA for multi-objective optimization. The fitness of each chromosome must be taken into account for all objective functions. Thus, algorithms for selecting the preferred chromosomes are needed. We focus on the selection phase and the fitness computation. To do that, we use and compare three strategies from the literature. They are described below.

1. **Fast nondominated sorting and crowding distance assignment (S1)**
   The fast nondominated sorting and crowding distance assignment methods were proposed for the Nondominated Sorting Genetic Algorithm II (NSGA-II) [5]. In our work, the fast nondominated sorting method is applied to classify the solutions in the population into several ranks. We only keep non dominated solutions. The number of nondominated solutions is compared to the elite set maximum size. If the number of nondominated solutions is less than or equal to this maximum size, all nondominated solutions will become the elite set. Otherwise, the crowding distance assignment method is applied to select some solutions from the nondominated set. The solutions in the boundary points of the nondominated set, which obtain the highest and lowest objective values, are always selected. Then,

the crowding distance values of all remaining solutions are computed and the $p_e - 2$ solutions, which have the highest crowding distance values, are selected. These solutions are included to the two pre-selected solutions from the boundary points, and then they are the members of the elite set in BRKGA process. If the elite set is not full, we consider the solutions of rank 2, i.e., the solutions only dominated by the solution of rank 1. The process is iterated until the elite set is full.

2. $\mathscr{S}$ *metric selection* **evolutionary multi-objective optimization algorithm (S2)**
   The $\mathscr{S}$ *metric selection* evolutionary multi-objective optimization algorithm or SMS-EMOA, which was proposed in [2], is applied to select some solutions in the current population to be added to the elite set. We compute the hypervolume defined by the non-dominated solutions in the population. Then, the population with the least contribution to the hypervolume are discarded until the bound on the size of the elite set is respected.

3. **Indicator-based evolutionary algorithm based on the hypervolume concept (S3)**
   The use of an indicator based on the hypervolume concept was proposed in the Indicator-Based Evolutionary Algorithm (IBEA) [14]. The indicator, namely the hypervolume, is used to assign fitness values to the population members. Then, some solutions in the current population are selected to be added to the elite set for the next population. The indicator based method performs binary tournaments for all solutions in the current population. The selection is implemented by removing the worst solution from the population and updating the fitness values of the remaining solutions. The worst solution is removed repeatedly until the number of remaining solutions satisfies the recommended size of the elite set for BRKGA.

# 3 Decoding Methods

In this section, the decoding methods, which are used for obtaining feasible solutions from random key chromosomes, are described. A chromosome consists of several genes. Each gene represents one acquisition, which needs to be scheduled. When the generation phase is over, the chromosome is decoded in order to obtain a sequence of acquisitions. The acquisitions are considered in an order deriving from a priority based on the associated gene value. The acquisition, which has the highest priority, will be firstly considered to be assigned in a sequence. Then, the next acquisitions are considered according to the priority order. The considered acquisition can be scheduled in the sequence only if all constraints are satisfied. Three methods for assigning the priority are studied in this paper. The decoder itself is a greedy heuristic that inserts at the end of the sequence the acquisition under consideration only if it respects all the operational constraints. The flowchart of constraint checking and acquisition assignment is depicted in Fig. 4. The example of one solution from the instance, which consists of two strips, is shown in Fig. 5. As previously presented, each strip can be considered as two acquisitions, according to the two possible acquiring directions,

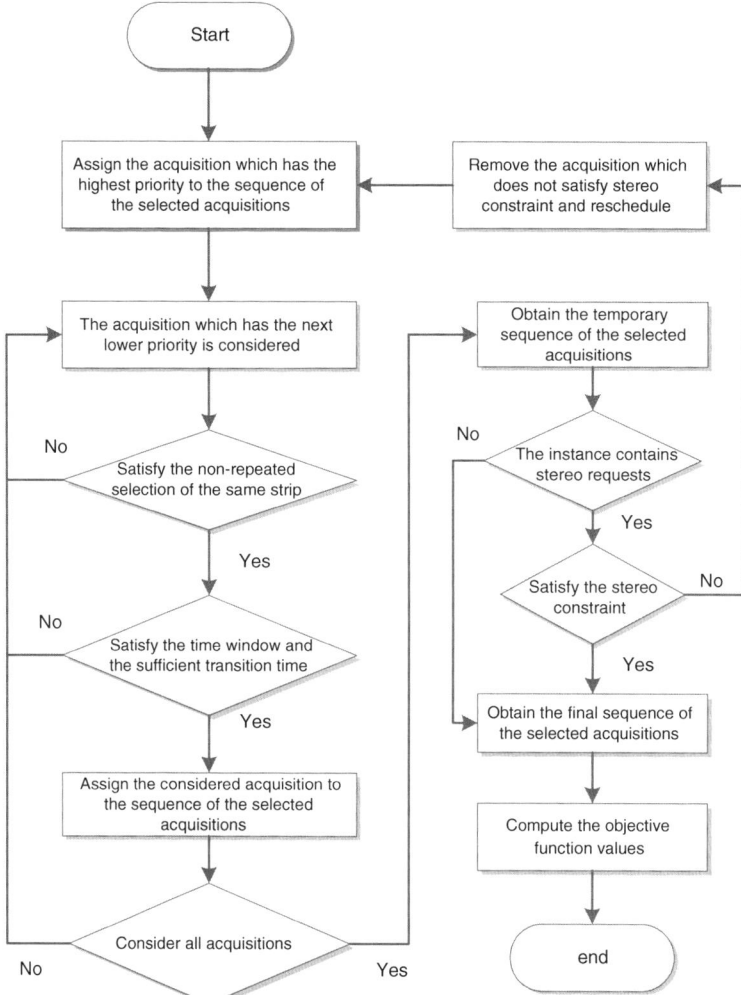

**Fig. 4** Flowchart of constraint checking and acquisition assignment

| Random-key chromosome | Acquisition0 | Acquisition1 | Acquisition2 | Acquisition3 |
|---|---|---|---|---|
| | 0.6984 | 0.9939 | 0.6885 | 0.2509 |
| Sequence of the selected acquisitions | 1   2 | | | |

**Fig. 5** Solution example from the modified instance, which needs to schedule two strips

but only one of them can be selected. Hence in this example, the size of random-key chromosome is equal to four, where a gene is associated with an acquisition. The figure shows the solution, which is decoded from the basic decoding (the priority to select and schedule of each acquisition equals to its gene value). The decoding step is used to obtain the sequence of the selected acquisition.

## 3.1 Basic Decoding (D1)

The first decoding method is a basic decoding. The priority of an acquisition $j$ is defined by using directly the gene value:

$$Priority_j = gene_j \qquad (1)$$

## 3.2 Decoding of Gene Value and Ideal Priority Combination (D2)

This decoding is presented in [11] to solve the resource-constrained project schedul-ing problem with makespan minimization. It considers the priority depending on the gene value, and also an ideal priority. For the concept of ideal priority, the job that has the earliest possible starting time should be selected first and be scheduled at the beginning of the sequence. Hence the ideal priority gives a higher priority to select and schedule the job with the earliest possible starting time. This ideal priority is a real value in the interval [0, 1] which is given by

$$\frac{LLP_j}{LCP}, \qquad (2)$$

where $LLP_j$ is the longest length path from the beginning of job $j$ to the end of the project and $LCP$ is the length along the critical path of the project.

The factor that adjusts the priority to account for the gene values of the random key chromosome is given by $(1 + gene_j)/2$. Thus, the second priority expression of each job $j$ is

$$Priority_j = \frac{LLP_j}{LCP} \times \left[ \frac{1 + gene_j}{2} \right] \qquad (3)$$

In the case of our problem, the priority expression of an acquisition $j$ becomes:

$$Priority_j = \frac{Tmax_L - Tmin_j}{Tmax_L} \times \left[ \frac{1 + gene_j}{2} \right] \qquad (4)$$

where $Tmax_L$ is the latest starting time of the last possible acquisition and $Tmin_j$ is the earliest starting time of acquisition $j$.

## 3.3 Hybrid Decoding (HD)

Finally, we propose a third decoding method, which is a hybrid method. It combines together the first and the second decoding methods. This hybrid method obtains two solutions from one chromosome, one for D1 and one for D2. Strategies must be designed to consider this fact, especially if the two solutions do not dominated each other. Three management schemes have been proposed and are described below.

### 3.3.1 Elite Set Management—Method 1 (M1)

Both solutions, obtained by the two decodings, are compared by using the dominance relation in the Pareto sense. If one solution dominates the other one, the dominant solution is kept, the other one discarded. Otherwise, one of the two solutions is selected randomly. The decoding process is repeated until all chromosomes in the population are decoded. When it finishes, the size of the solution set is equal to $p$. Then, $p_e$ solutions are selected to become the elite set by using the same methods with only one decoding. The principle of elite set management by Method 1 is shown in Fig. 6.

### 3.3.2 Elite Set Management—Method 2 (M2)

All chromosomes in the population are decoded by using the two decoding methods. Two solutions are obtained from the decoding of one chromosome. Both of them are stored in the solution set. Hence, the size of the solution set is equal to $2p$. Then, $p_e$ solutions are selected from the solution set to become the elite set using their fitness (as explained earlier). The principle of elite set management by Method 2 is shown in Fig. 7.

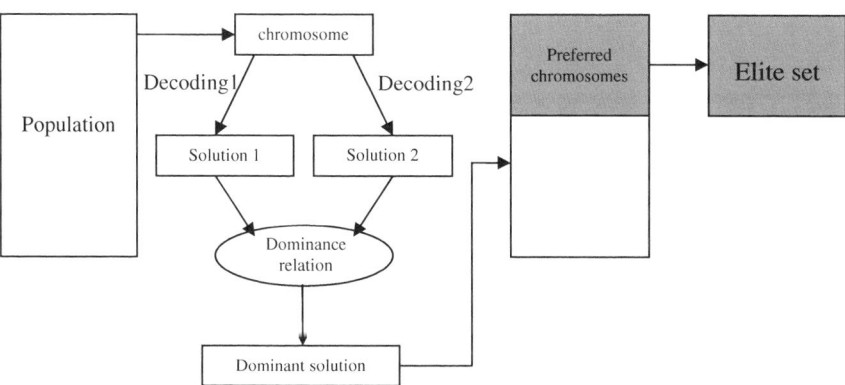

**Fig. 6** Elite set management for hybrid decoding—Method 1

**Fig. 7** Elite set management
for hybrid
decoding—Method 2

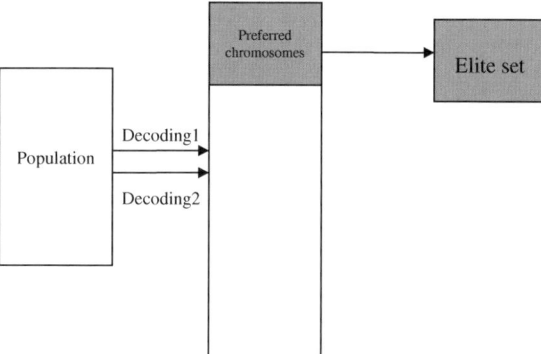

**Fig. 8** Elite set management
for hybrid
decoding—Method 3

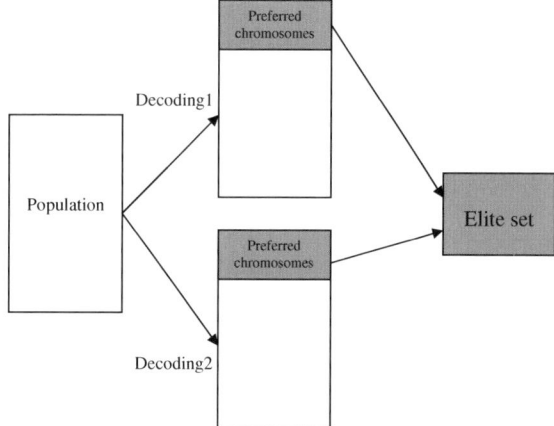

### 3.3.3 Elite Set Management—Method 3 (M3)

Each chromosome in the population is firstly decoded by using the priority equation
of basic decoding, and the obtained solution is stored in a first solution set. Similarly,
the same chromosome is decoded by using D2 and the obtained solution is stored in
a second solution set. When all chromosomes in the population are decoded and the
solutions are stored into two solution sets, the selection methods are applied to select
$p_e$ solutions for becoming the elite set. Hence, the $p_e/2$ preferred solutions must be
chosen from each solution set, as shown in Fig. 8.

## 4 Computational Results

The methods are tested on modified ROADEF 2003 challenge instances (Testset A)
to consider four users. The format of instance names are changed to a_b_c, where
a is the number of requests, b is the number of stereo requests, and c is the number
of strips.

For the proposed biased random-key genetic algorithm (BRKGA), parameter values of the algorithm were experimentally tuned for our work. The population size of BRKGA is set equal to the length of the random-key chromosome or twice the number of strips. The sizes of the three parts, which are generated to become the population in the next generation, are set in accordance with the recommended values in Table 1. The size of the elite set is equal to the number of non-repeating schedules from the nondominated solutions, but it is not over $0.15p$. The size of mutant set is equal to $0.3p$. The probability of elite element inheritance for crossover operation is set to 0.6. In each iteration, the nondominated solutions are stored in an archive. If there is at least one solution from the current population that can dominate some solutions in the archive, the archive will be updated. Therefore, we use the number of iterations since the last archive improvement to be a stopping criterion. We opt for 50 iterations. Moreover, the computation time is used to be the second stopping criterion. It is adapted to the instance size. The iteration of BRKGA will be stopped, when one of the two stopping criteria is satisfied. The algorithm is implemented in C++ and thirty runs per instance are tested. Tests were performed on a dual core Intel x86-64 Xeon processor W3520 (2.67 GHz) with 6 GB RAM under GNU/Linux 3.2.0-58-generic. The hypervolumes of the approximate Pareto front are computed by using a reference point of 0 for the first objective (maximizing the total profit) and the maximum of the profit summations of each user for the second one (minimizing the profit difference between users). Three elite selecting methods from three efficient algorithms: NSGA-II, SMS-EMOA, and IBEA, are applied to select some solutions to become the elite set. The set of testing instances consists of ten instances. However, the results of the smallest instance (instance 2_0_2) cannot be reached, when using the population size equal to the length of the chromosome or twice of number of strips, because the population size is too small for generating the new population from the three sets of chromosomes in BRKGA process. Hence, the results of nine instances will be presented in the experimental results.

Firstly, the three elite selection methods (S1–S3) are tested and the basic decoding is used to decode the chromosomes to become the solutions. The results of each elite selection methods are compared. The hypervolume values of the approximate Pareto front are computed. Box plots, which illustrate the maximum value, the median value, the minimum value, and the interquartile range of the hypervolume values, and the average computation times are presented in Fig. 9. For each instance, the first, second, and third columns illustrate the hypervolume values from the elite selection methods borrowed from NSGA-II (S1), SMS-EMOA (S2), and IBEA (S3), respectively. Moreover, we also use a Mann-Whitney statistical test [9] for comparing the hypervolume values as shown in Table 2 and the computation times in Table 3. The results show that each selection method has advantages in different instances regarding the hypervolume values. However, Table 3 shows that S3 spends less computation time for most instances. Therefore, only the selection method S3 will be used to select the preferred chromosomes in the next experiments.

Secondly, the three methods of elite set management for hybrid decoding (M1–M3) are tested by using the elite set selection method borrowed from IBEA. The hypervolume values of the approximate Pareto front, which are obtained from each

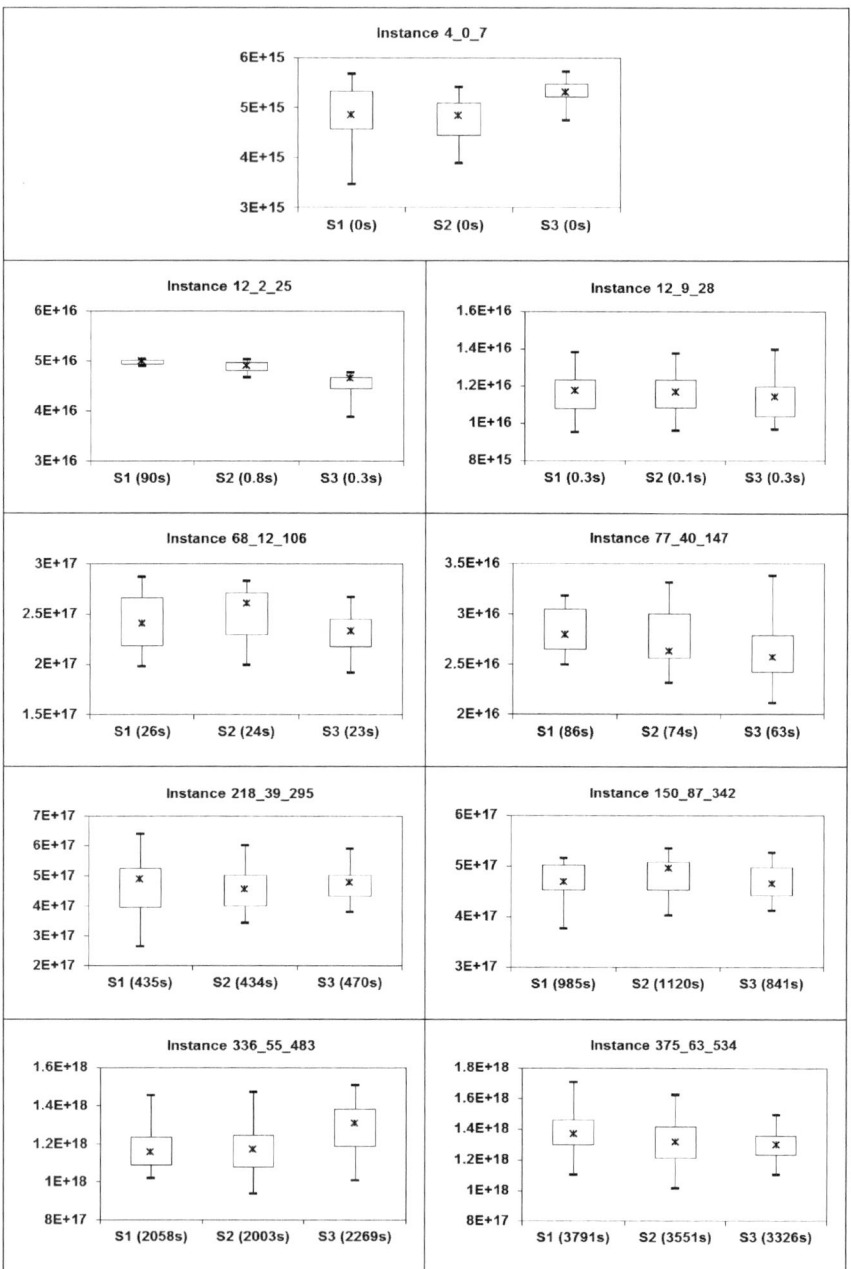

**Fig. 9** Comparison of the results of the three elite selection methods by using the basic decoding (hypervolume values and CPU times)

**Table 2** Mann-Whitney statistical test of the hypervolume values from the three elite selection methods ($\succ$, $\prec$ respectively stand for statistically better, worse)

| Inst. 4_0_7 | S1 | S2 | S3 | Inst. 12_2_25 | S1 | S2 | S3 | Inst. 12_9_28 | S1 | S2 | S3 |
|---|---|---|---|---|---|---|---|---|---|---|---|
| S1 | − | $\succ$ | $\prec$ | S1 | − | $\succ$ | $\succ$ | S1 | − | $\prec$ | $\succ$ |
| S2 | $\prec$ | − | $\prec$ | S2 | $\prec$ | − | $\succ$ | S2 | $\succ$ | − | $\succ$ |
| S3 | $\succ$ | $\succ$ | − | S3 | $\prec$ | $\prec$ | − | S3 | $\prec$ | $\prec$ | − |
| Inst. 68_12_106 | S1 | S2 | S3 | Inst. 77_40_147 | S1 | S2 | S3 | Inst. 218_39_295 | S1 | S2 | S3 |
| S1 | − | $\prec$ | $\succ$ | S1 | − | $\succ$ | $\succ$ | S1 | − | $\succ$ | $\prec$ |
| S2 | $\succ$ | − | $\succ$ | S2 | $\prec$ | − | $\succ$ | S2 | $\prec$ | − | $\prec$ |
| S3 | $\prec$ | $\prec$ | − | S3 | $\prec$ | $\prec$ | − | S3 | $\succ$ | $\succ$ | − |
| Inst. 150_87_342 | S1 | S2 | S3 | Inst. 336_55_483 | S1 | S2 | S3 | Inst. 375_63_534 | S1 | S2 | S3 |
| S1 | − | $\prec$ | $\succ$ | S1 | − | $\succ$ | $\prec$ | S1 | − | $\succ$ | $\succ$ |
| S2 | $\succ$ | − | $\succ$ | S2 | $\prec$ | − | $\prec$ | S2 | $\prec$ | − | $\succ$ |
| S3 | $\prec$ | $\prec$ | − | S3 | $\succ$ | $\succ$ | − | S3 | $\prec$ | $\prec$ | − |

**Table 3** Mann-Whitney statistical test of the computation times from the three elite selection methods ($\prec$, $\succ$, $\equiv$ respectively stand for statistically better, worse, identical)

| Inst. 4_0_7 | S1 | S2 | S3 | Inst.12_2_25 | S1 | S2 | S3 | Inst. 12_9_28 | S1 | S2 | S3 |
|---|---|---|---|---|---|---|---|---|---|---|---|
| S1 | − | $\equiv$ | $\equiv$ | S1 | − | $\succ$ | $\succ$ | S1 | − | $\succ$ | $\prec$ |
| S2 | $\equiv$ | − | $\equiv$ | S2 | $\prec$ | − | $\succ$ | S2 | $\prec$ | − | $\prec$ |
| S3 | $\equiv$ | $\equiv$ | − | S3 | $\prec$ | $\prec$ | − | S3 | $\succ$ | $\succ$ | − |
| Inst. 68_12_106 | S1 | S2 | S3 | Inst. 77_40_147 | S1 | S2 | S3 | Inst. 218_39_295 | S1 | S2 | S3 |
| S1 | − | $\succ$ | $\succ$ | S1 | − | $\succ$ | $\succ$ | S1 | − | $\equiv$ | $\prec$ |
| S2 | $\prec$ | − | $\succ$ | S2 | $\prec$ | − | $\succ$ | S2 | $\equiv$ | − | $\prec$ |
| S3 | $\prec$ | $\prec$ | − | S3 | $\prec$ | $\prec$ | − | S3 | $\succ$ | $\succ$ | − |
| Inst. 150_87_342 | S1 | S2 | S3 | Inst. 336_55_483 | S1 | S2 | S3 | Inst. 375_63_534 | S1 | S2 | S3 |
| S1 | − | $\prec$ | $\succ$ | S1 | − | $\succ$ | $\prec$ | S1 | − | $\equiv$ | $\succ$ |
| S2 | $\succ$ | − | $\succ$ | S2 | $\prec$ | − | $\prec$ | S2 | $\equiv$ | − | $\succ$ |
| S3 | $\prec$ | $\prec$ | − | S3 | $\succ$ | $\succ$ | − | S3 | $\prec$ | $\prec$ | − |

elite set management method, are displayed in box plot and the average computation times are reported in Fig. 10. The results show that the three methods obtain similar solutions regarding the hypervolume values.

Therefore, we use a Mann-Whitney statistical test for comparing the hypervolume values, as shown in Table 4, and the computation times in Table 5. Each method obtains better hypervolume values for different instances. Thus, we compare each elite set management method by considering the computation times. The statistical results of the computation times show that M2 and M3 spend much higher computation time than M1 for all instances. Therefore in the sequel, only method M1 will be kept to compare the results between the hybrid decoding (HD–M1) and the two single ones (D1 and D2).

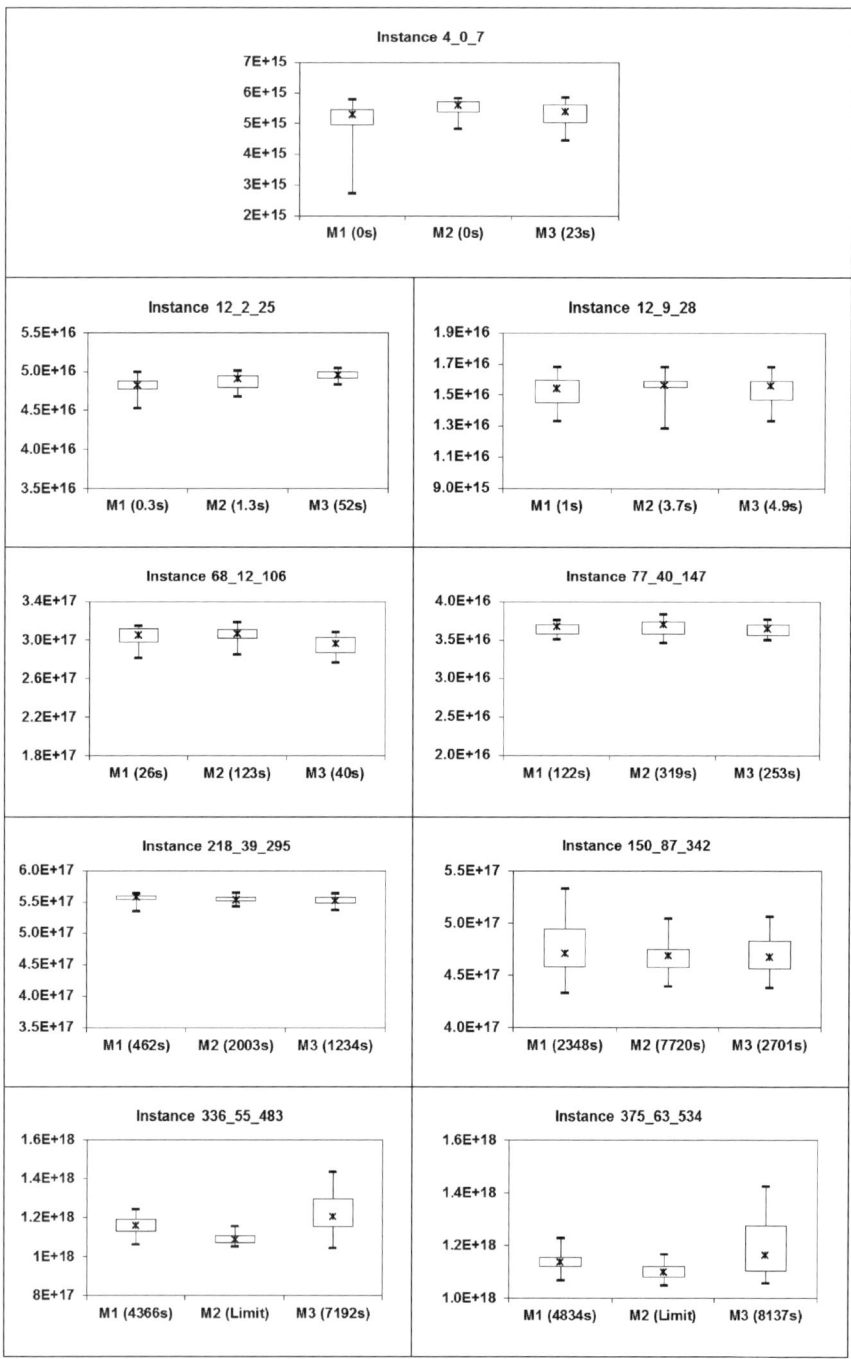

**Fig. 10** Comparison of the results of the three methods for management of the elite set for hybrid decoding (M1, M2, and M3) by using the method for elite set selection borrowed from IBEA (hypervolume values and CPU times)

**Table 4** Mann-Whitney statistical test of the hypervolume values from the three elite set management methods for the hybrid decoding ($\succ$, $\prec$, $\equiv$ respectively stand for statistically better, worse, identical)

| Inst. 4_0_7 | M1 | M2 | M3 | Inst. 12_2_25 | M1 | M2 | M3 | Inst. 12_9_28 | M1 | M2 | M3 |
|---|---|---|---|---|---|---|---|---|---|---|---|
| M1 | − | ≺ | ≺ | M1 | − | ≺ | ≺ | M1 | − | ≺ | ≡ |
| M2 | ≻ | − | ≻ | M2 | ≻ | − | ≺ | M2 | ≻ | − | ≻ |
| M3 | ≻ | ≺ | − | M3 | ≻ | ≻ | − | M3 | ≡ | ≺ | − |
| Inst. 68_12_106 | M1 | M2 | M3 | Inst. 77_40_147 | M1 | M2 | M3 | Inst. 218_39_295 | M1 | M2 | M3 |
| M1 | − | ≺ | ≻ | M1 | − | ≺ | ≻ | M1 | − | ≻ | ≻ |
| M2 | ≻ | − | ≻ | M2 | ≻ | − | ≻ | M2 | ≺ | − | ≻ |
| M3 | ≺ | ≺ | − | M3 | ≺ | ≺ | − | M3 | ≺ | ≺ | − |
| Inst. 150_87_342 | M1 | M2 | M3 | Inst. 336_55_483 | M1 | M2 | M3 | Inst. 375_63_534 | M1 | M2 | M3 |
| M1 | − | ≻ | ≻ | M1 | − | ≻ | ≺ | M1 | − | ≻ | ≺ |
| M2 | ≺ | − | ≡ | M2 | ≺ | − | ≺ | M2 | ≺ | − | ≺ |
| M3 | ≺ | ≡ | − | M3 | ≻ | ≻ | − | M3 | ≻ | ≻ | − |

**Table 5** Mann-Whitney statistical test of the computation times from the three elite set management methods for the hybrid decoding ($\prec$, $\succ$, $\equiv$ respectively stand for statistically better, worse, identical)

| Inst. 4_0_7 | M1 | M2 | M3 | Inst. 12_2_25 | M1 | M2 | M3 | Inst. 12_9_28 | M1 | M2 | M3 |
|---|---|---|---|---|---|---|---|---|---|---|---|
| M1 | − | ≡ | ≺ | M1 | − | ≺ | ≺ | M1 | − | ≺ | ≺ |
| M2 | ≡ | − | ≺ | M2 | ≻ | − | ≺ | M2 | ≻ | − | ≻ |
| M3 | ≻ | ≻ | − | M3 | ≻ | ≻ | − | M3 | ≻ | ≺ | − |
| Inst. 68_12_106 | M1 | M2 | M3 | Inst. 77_40_147 | M1 | M2 | M3 | Inst. 218_39_295 | M1 | M2 | M3 |
| M1 | − | ≺ | ≺ | M1 | − | ≺ | ≺ | M1 | − | ≺ | ≺ |
| M2 | ≻ | − | ≻ | M2 | ≻ | − | ≻ | M2 | ≻ | − | ≻ |
| M3 | ≻ | ≺ | − | M3 | ≻ | ≺ | − | M3 | ≻ | ≺ | − |
| Inst. 150_87_342 | M1 | M2 | M3 | Inst. 336_55_483 | M1 | M2 | M3 | Inst. 375_63_534 | M1 | M2 | M3 |
| M1 | − | ≺ | ≺ | M1 | − | ≺ | ≺ | M1 | − | ≺ | ≺ |
| M2 | ≻ | − | ≻ | M2 | ≻ | − | ≻ | M2 | ≻ | − | ≻ |
| M3 | ≻ | ≺ | − | M3 | ≻ | ≺ | − | M3 | ≻ | ≺ | − |

Finally, the three decoding methods (D1, D2, and HD) are tested and the obtained results are compared. The box plots from the elite set selection method, which borrowed from IBEA, are reported in Fig. 11. The graph illustrates the box plots of the hypervolume values, and the average computation times are presented.

Most of the results show that the hybrid decoding obtains results close to the best ones, when comparing the two single decodings, although it does not obtain certainly the best results in some instances. Indeed, it can preserve the advantages of the two single decodings for all instances. For example, in instance 12_2_25, the first decoding method obtains better results than the second one, thus the hybrid decoding obtains results similar to the first one. For instance 77_40_147, the hybrid

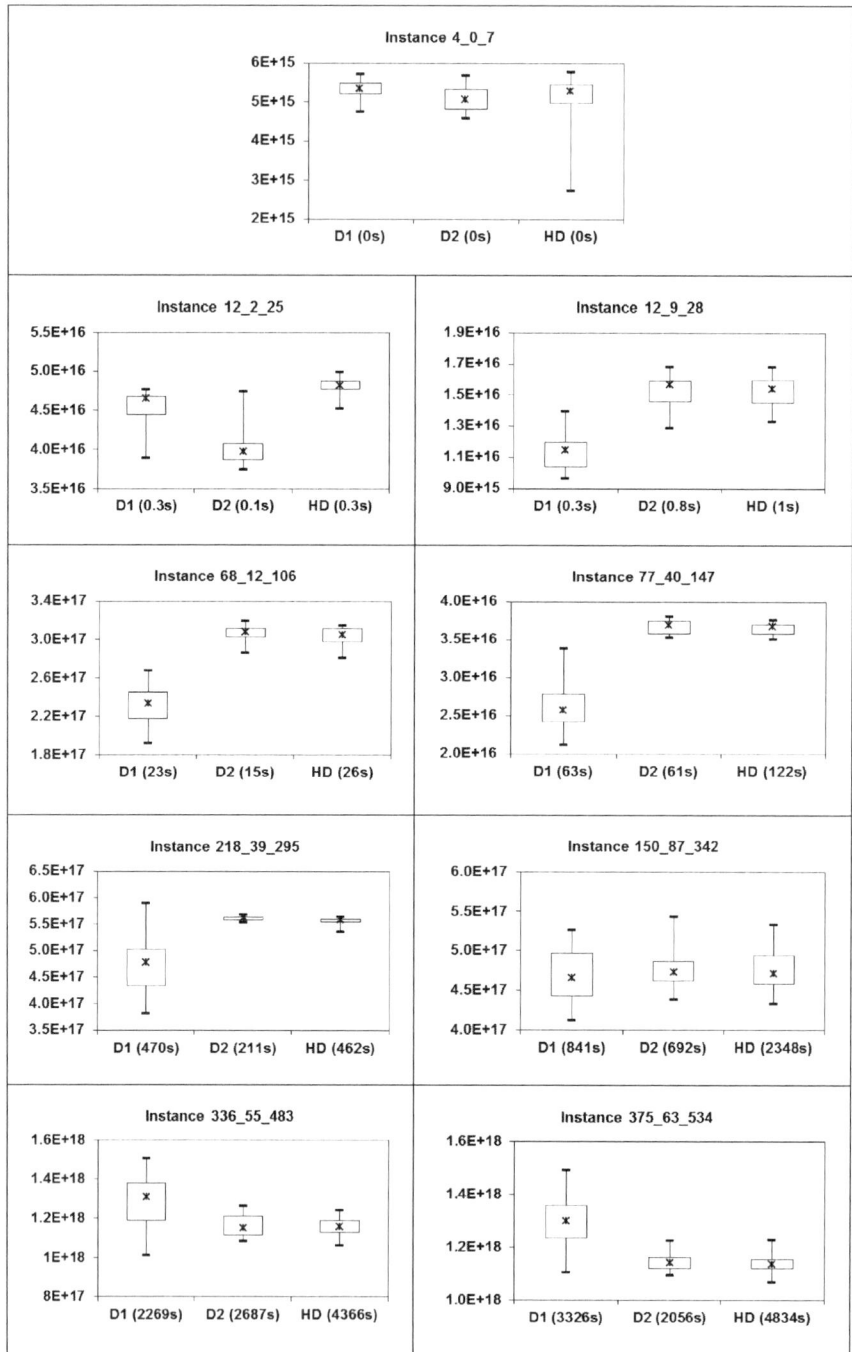

**Fig. 11** Comparison of the results of the three decoding methods (D1, D2, and HD–M1) by using the method for elite set selection borrowed from IBEA (hypervolume values and CPU times)

decoding obtains results similar to the second decoding, which obtains better results than the first one. Thus, the hybrid decoding method is efficient for solving most instances. Compared with D1, it can reduce the range of hypervolume values. This means that the hybrid decoding can provide results with better standard deviations. Moreover, for some instances where the second decoding entraps in local optima, the hybrid decoding is able to reach the global optimum. Regarding the computation time, the hybrid decoding method spends longer time in each iteration, however it can obtain good solutions in a reasonable computation time, which is limited by the second stopping criterion of BRKGA process.

# 5  Conclusions and Perspectives

A biased random-key genetic algorithm or BRKGA is used for solving a multi-objective optimization problem. The BRKGA works on a chromosome encoded as a key vector. The chromosome consists of several genes, which are encoded by real values in the interval $[0, 1]$. For applying to the multi-objective optimization problem, the elite set selection method has to be defined in the BRKGA process. Moreover, during each iteration of BRKGA, the chromosomes are decoded to obtain the feasible solutions. A hybrid decoding, which combines two single decodings, is proposed in this paper. Two solutions are obtained from the decoding of one chromosome, when using the hybrid decoding. Thus, the methods for elite set management have to be defined and three methods are tested.

The experiments are conducted on the multi-user observation scheduling problem for agile Earth observing satellites. The requests are required from multiple users. The objectives of this problem are to maximize the total profit and simultaneously minimize the maximum profit difference between users to ensure the sharing fairness. Three elite selecting methods, which are borrowed from NSGA-II, SMS-EMOA, and IBEA, are used for selecting a set of preferred solutions to become the elite set of the population. For the three elite selecting methods, the hypervolume values and the average computation times are compared. Then, the hybrid decoding is proposed and its results are compared with the results from the two single decodings. The hybrid decoding can preserve the advantages of the two single decodings, since it obtains results close to the best results of the two single decodings in reasonable computation times. Moreover, it can improve the standard deviation of the hypervolume values and avoid to entrap in local optima. Finally, the hybrid decoding is proper to be applied in BRKGA process for solving multi-objective optimization problems, which need several feasible solutions on the Pareto front.

As perspectives, we should investigate some further works. First, the hybrid decoding step and the elite set selection method could be modified. The two single decodings that were used and combined to be the hybrid decoding, could also be changed. Indeed, each of these single decodings may be re-defined to mainly consider one objective. Then, for the elite set selection, other indicators could be used like the selection method borrowed from SPEA2. Moreover, more advanced decoding methods could also be applied for BRKGA, e.g. considering the decoder as a full multi-objective problem.

# References

1. Bean, J.C.: Genetic algorithms and random keys for sequencing and optimization. ORSA J. Comput. **6**, 154–160 (1994)
2. Beume, N., Naujoks, B., Emmerich, M.: SMS-EMOA: Multiobjective selection based on dominated hypervolume. Eur. J. Oper. Res. **181**, 1653–1669 (2007)
3. Bianchessi, N., Cordeau, J.F., Desrosiers, J., Laporte, G., Raymond, V.: A heuristic for the multi-satellite, multi-orbit and multi-user management of earth observation satellites. Eur. J. Oper. Res. **177**, 750–762 (2007)
4. Cordeau, J.F., Laporte, G.: Maximizing the value of an earth observation satellite orbit. J. Oper. Res. Soc. **56**, 962–968 (2005)
5. Deb, K., Pratep, A., Agarwal, S., Meyarivan, T.: A fast and elite multiobjective genetic algorithm: NSGA-II. IEEE Trans. Evol. Comput. **6**, 182–197 (2002)
6. Gonçalves, J.F., Almeida, J.: A hybrid genetic algorithm for assembly line balancing. J. Heuristics **8**, 629–642 (2002)
7. Gonçalves, J.F., Resende, M.G.C.: Biased random-key genetic algorithms for combinatorial optimization. J. Heuristics **17**, 487–525 (2011)
8. Goulart, N., de Souza, S. R., Dias, L. G. S., Noronha, T. F.: Biased Random-key Genetic Algorithm for Fiber Installation in Optical Network Optimization. In: IEEE Congress on Evolutionary Computation, pp. 2267–2271. New Orleans (2011)
9. Knowles, J., Thiele, L., Zitzler, E.: Technical report, Computer Engineering and Networks Laboratory (TIK). A tutorial on the performance assessment of stochastic multiobjective optimizers. ETH Zurich, Switzerland (2006)
10. Kuipers, E. J.: An Algorithm for Selecting and Timetabling Requests for an Earth Observation Satellite. Bulletin de la Société Française de Recherche Opérationnelle et d'Aide à la Décision, pp. 7–10 (2003) (available at: http://www.roadef.org/content/roadef/bulletins/bulletinNo11.pdf)
11. Mendes, J.J.M., Gonçalves, J.F., Resende, M.G.C.: A random key based genetic algorithm for the resource constrained project scheduling problem. Comput. Oper. Res. **36**, 92–109 (2009)
12. Tangpattanakul, P., Jozefowiez, N., Lopez, P.: Multi-objective Optimization for Selecting and Scheduling Observations by Agile Earth Observing Satellites. In: Coello Coello, C., Cutello, V., Deb, K., Forrest, S., Nicosia, G., Pavone, M. (eds.) PPSN XII. LNCS, vol. 7492, pp. 112–121. Springer, Heidelberg (2012)
13. Verfaillie, G., Lemaître, M., Bataille, N., Lachiver, J. M.: Management of the mission of earth observation satellites challenge description. Technical report, Centre National d'Etudes Spatiales, France (2002) (available at: http://challenge.roadef.org/2003/files/formal_250902.pdf)
14. Zitzler, E., Künzli, S.: Indicator-Based selection in multiobjective search. In: Yao, X., Burke, E.K., Lozano, J.A., Smith, J., Merelo-Guervós, J.J., Bullinaria, J.A., Rowe, J.E., Tiňo, P., Kabán, A., Schwefel, H.-P. (eds.) PPSN VIII. LNCS, vol. 3242, pp. 832–842. Springer, Heidelberg (2004)

# Efficient and Scalable Pareto Front Generation for Energy and Makespan in Heterogeneous Computing Systems

Kyle M. Tarplee, Ryan Friese, Anthony A. Maciejewski
and Howard Jay Siegel

**Abstract** The rising costs and demand of electricity for high-performance computing systems pose difficult challenges to system administrators that are trying to simultaneously reduce operating costs and offer state-of-the-art performance. However, system performance and energy consumption are often conflicting objectives. Algorithms are necessary to help system administrators gain insight into this energy/performance trade-off. Through the use of intelligent resource allocation techniques, system administrators can examine this trade-off space to quantify how much a given performance level will cost in electricity, or see what kind of performance can be expected when given an energy budget. A novel algorithm is presented that efficiently computes tight lower bounds and high quality solutions for energy and makespan. These solutions are used to bound the Pareto front to easily trade-off energy and performance. These new algorithms are shown to be highly scalable in terms of solution quality and computation time compared to existing algorithms.

**Keywords** High performance computing · Scheduling · Bag-of-tasks · Scalable · Efficient · Heterogeneous computing

K.M. Tarplee (✉) · R. Friese · A.A. Maciejewski · H.J. Siegel
Department of Electrical and Computer Engineering,
Colorado State University, Fort Collins, CO 80523, USA
e-mail: kyle.tarplee@colostate.edu; ktarplee@me.com

R. Friese
e-mail: ryan.friese@colostate.edu

A.A. Maciejewski
e-mail: aam@colostate.edu

H.J. Siegel
Department of Computer Science, Colorado State University,
Fort Collins, CO 80523, USA
e-mail: hj@colostate.edu

© Springer International Publishing Switzerland 2015    161
S. Fidanova (ed.), *Recent Advances in Computational Optimization*,
Studies in Computational Intelligence 580, DOI 10.1007/978-3-319-12631-9_10

# 1 Introduction

The race for increased performance in high-performance computing (HPC) systems has resulted in a large increase in the power consumption of these systems [1]. This increase in power consumption can cause degradation in the electrical infrastructure that supports these facilities, as well as increase electricity costs for the operators [2]. The goals of HPC users conflict with the HPC operators in that the users' goal is to finish their workload as quickly as possible. That is, the small energy consumption desired by the system operator and the high system performance desired by the users are conflicting objectives that require the sacrifice of one to improve the other. Balancing the performance needs of the users with energy costs proves difficult without tools designed to help a system administrator choose from among a set of solutions.

A set of efficient and scalable algorithms are proposed that can help system administrators quickly gain insight into the energy and performance trade-off of their HPC systems through the use of intelligent resource allocation. The algorithms proposed have very desirable run times and produce schedules that are closer to optimal as the problem size increases. As such, this approach is very well suited to large scale HPC systems.

The focus of our work is on a common scheduling problem where the users submit a set of independent tasks known as a *bag-of-tasks* [3]. The tasks will run on a dedicated set of interconnected machines. A task runs on only one machine and, likewise, a machine may only process one task at any one time. This class of scheduling problems is often referred to as *static scheduling* because the full bag-of-tasks is known a priori [4]. Task execution and power consumption are deterministic in this model. The HPC systems of primary interest have highly heterogeneous task run times, machines, and power consumption which are known as *heterogeneous computing* (HC) systems. Some machines in the HC systems are often special purpose machines that can perform specific tasks quickly, while other tasks might not be able to run at all on that hardware. Another cause of heterogeneity is differing computational requirements, input/output bottlenecks, or memory limitations, and therefore cannot take full advantage of the machine. The machines may further differ in the average power consumed for each task type. Machines may have different architectures, leading to vastly different power consumption characteristics. For instance, a task that runs on a GPU might consume less energy to complete, but often more power, than the same task run on a general purpose machine, due to the shorter execution time. We assume one objective is to minimize the maximum finishing time of all tasks, which is known as the *makespan*. The heterogeneity in execution time of the tasks provides the scheduler degrees of freedom to greatly improve the makespan over a naïve scheduling algorithm. Similarly the heterogeneity in the power consumption allows the schedulers to decrease the energy consumption.

The contributions of this paper are:

1. The formulation of an algorithm that efficiently computes tight lower bounds on the energy and makespan and quickly recovers near optimal feasible solutions.

2. Finding a high quality bi-objective Pareto front.
3. An evaluation of the scaling properties of the proposed algorithms.
4. The addition of idle power consumption to the formulation of the energy/makespan problem in [3].

The rest of this paper is as follows: first the lower bound on the objectives is described in Sect. 2.2. Then algorithms are presented in Sects. 2.3–2.5 that reconstruct a feasible schedule from the lower bound. In Sect. 2.6, the complexity of the algorithm is analyzed. Algorithm scaling quality and runtime results are shown in Sect. 3. Section 4 shows how the lower bounds can be used with any scalarization technique to form a Pareto front. Section 5 compares these algorithms to the NSGA-II algorithm. Section 6 discusses related work and Sect. 7 concludes while presenting some ideas for future work.

## 2 Approximation Algorithms

### 2.1 Approach

The fundamental approach of this paper is to apply *divisible load theory* (DLT) [5] to ease the computational requirements of computing a lower bound solution on the energy and makespan. For the lower bound, a single task is allowed to be divided and scheduled onto any number of machines. After the lower bound on the energy and makespan is computed, a two phase algorithm is used to recover a feasible solution from the infeasible lower bound solution. The feasible solution serves as the upper bound on the optimal energy and makespan.

Often HC systems have groups of machines, usually purchased at the same time, that have identical or nearly identical performance and power characteristics. Even when every machine is different, the uncertainty in the system often allows one to model similar machines as groups of machines of the same type. Machines that have virtually indistinguishable performance and power properties with respect to the workload are said to be the same *machine type*. Machines within a machine type may differ vastly in feature sets so long as the task performance and power consumption of the tasks under consideration are not affected. Tasks often exhibit natural groupings as well. Tasks of the same *task type* are often submitted many times to perform statistical simulations and other repetitive jobs. In fact, having groupings for tasks and for machines permits less profiling effort to estimate the run time and power consumption for each task on each machine.

Traditionally this static scheduling problem is posed as assigning all tasks to all machines. The classic formulation is not well suited for recovering a high quality feasible solution. The decision variables would be binary valued (assigned or not assigned) and rounding a real value from the lower bound to a binary value can change the objective significantly. Complicated rounding schemes are necessary to iteratively compute a suitable solution. Instead, the problem is posed as determining the number of tasks of *each type* to assign to each *machine type*. With this modification, decision variables will be large integers $\gg 1$, resulting in only a small error to the objective function when rounding to the nearest integer. This approximation holds well when the number of tasks assigned to each machine type is large. For this approximation, machine types need not be large. In addition to easing the recovery of the integer solution, another benefit of this formulation is that it is much less computationally intensive due to solving the higher level assignment of tasks types to machine types with DLT, before solving the fine grain assignment of individual tasks to machines. As such, this approach can be thought of as a hierarchical solution to the static scheduling problem.

## 2.2 Lower Bound

The lower bound is given by the solution to a linear bi-objective optimization (or vector optimization) problem and is constructed as follows. Let there be $T$ task types and $M$ machine types. Let $T_i$ be the number of tasks of type $i$ and $M_j$ be the number of machines of type $j$. Let $x_{ij}$ be the number of tasks of type $i$ assigned to machine type $j$, where $x_{ij}$ is the primary decision variable in the optimization problem. Let $ETC$ be a $T \times M$ matrix where $ETC_{ij}$ is the *estimated time to compute* for a task of type $i$ on a machine of type $j$. Similarly let $APC$ be a $T \times M$ matrix where $APC_{ij}$ is the *average power consumption* for a task of type $i$ on a machine of type $j$. These matrices are frequently used in scheduling algorithms [4, 6–8]. $ETC$ and $APC$ are generally determined empirically.

The lower bound of the finishing times of a machine type is found by allowing tasks to be divided among all machines to ensure the minimal finishing time. With this conservative approximation all tasks in machine type $j$ finish at the same time. The finishing time of machine type $j$, denoted by $F_j$ is given by

$$F_j = \frac{1}{M_j} \sum_i x_{ij} ETC_{ij}. \tag{1}$$

Throughout this work sums over $i$ always go from 1 to $T$ and sums over $j$ always go from 1 to $M$, thus the ranges are omitted. Given that $F_j$ is a lower bound on the finishing time for a machine type, the tightest lower bound on the makespan is:

$$MS_{LB} = \max_j F_j. \tag{2}$$

Energy consumed by the bag-of-tasks is $\sum_i \sum_j x_{ij} APC_{ij} ETC_{ij}$. To incorporate idle power consumption, one must consider the time duration for which the machines are powered on. In this model, the time duration is the makespan. Not all machines will finish executing tasks at the same time. All but the last machine(s) to finish will accumulate idle power. When no task is executing on machine $j$, the power consumption is given by the idle power consumption, $APC_{\emptyset j}$. The equation for the lower bound on the energy consumed, incorporating idle power, is given in:

$$
\begin{aligned}
E_{LB} &= \sum_i \sum_j x_{ij} APC_{ij} ETC_{ij} \\
&\quad + \sum_j M_j APC_{\emptyset j} (MS_{LB} - F_j) \\
&= \sum_i \sum_j x_{ij} ETC_{ij} \left( APC_{ij} - APC_{\emptyset j} \right) \\
&\quad + \sum_j M_j APC_{\emptyset j} MS_{LB}
\end{aligned}
\tag{3}
$$

where the second term in the first equation accounts for the idle power.

The resulting bi-objective optimization problem for the lower bound is:

$$
\begin{aligned}
\underset{x,\, MS_{LB}}{\text{minimize}} \quad & \begin{pmatrix} E_{LB} \\ MS_{LB} \end{pmatrix} \\
\text{subject to:} \quad \forall i \qquad & \sum_j x_{ij} = T_i \\
\forall j \qquad & F_j \leq MS_{LB} \\
\forall i, j \qquad & x_{ij} \geq 0.
\end{aligned}
\tag{4}
$$

The objective of Eq. (4) is to minimize $E_{LB}$ and $MS_{LB}$, where $x$ is the primary decision variable. $MS_{LB}$ is an auxiliary decision variable necessary to model the objective function in Eq. (2). The first constraint ensures that all tasks in the bag are assigned to some machine type. The second constraint is the makespan constraint. Because the objective is to minimize makespan, the $MS_{LB}$ variable will be equal to the maximum finishing time of all the machine types. The third constraint ensures that there are no negative assignments in the solutions. This vector optimization problem can be solved to find a collection of optimal solutions. It is often solved by weighting the objective functions to form a *linear programming* (LP) problem or *linear program*. Methods to find a collection of solutions are presented in Sect. 4.

Ideally this vector optimization problem would be solved optimally with $x_{ij} \in \mathbb{Z}_{\geq 0}$. However, for practical scheduling problems, finding the optimal integral solution is often not possible due to the high computational cost. Fortunately, efficient algorithms to produce high quality sub-optimal feasible solutions exist. The next few sections describe how to take an infeasible real-valued solution from the linear program and build a complete feasible allocation.

## 2.3 Recovery Algorithm

An algorithm is necessary to recover a feasible solution or full resource allocation from each infeasible solutions obtained from the lower bound solutions of Eq. (4). Numerous approaches have been proposed in the literature for solving integer LP problems by first relaxing them to real-valued LP problems [9]. The approach here follows this common technique combined with computationally inexpensive techniques tailored to this particular optimization problem. The recovery algorithm is decomposed into two phases. The first phase rounds the solution while taking care to maintain feasibility of Eq. (4). The second phase assigns tasks to actual machines to build the full task allocation. The next two sections detail the two phases of this recovery algorithm.

## 2.4 Rounding

Due to the nature of the problem, the optimal solution $x^*$ often has few non-zero elements per row. Usually all the tasks of one type will be assigned to a small number of machine types. In the original problem, tasks are not divisible so one needs to have an integer number of tasks to assign to a machine type. When tasks are split between machine types, an algorithm is needed to compute an integer solution from this real-valued solution. The following algorithm finds $\hat{x}_{ij} \in \mathbb{Z}_{\geq 0}$ such that it is near $x^*_{ij}$ while maintaining the task assignment constraint. Algorithm 1 finds $\hat{x}$ that minimizes $\| \hat{x}_{ij} - x^*_{ij} \|_1$ for a given $i$.

---

**Algorithm 1** Round to the nearest integer solution while maintaining the constraints

---

1: **for** $i = 1$ to $T$ **do**
2:    $n \leftarrow T_i - \sum_j \lfloor x^*_{ij} \rfloor$
3:    $\forall j \quad f_j \leftarrow x^*_{ij} - \lfloor x^*_{ij} \rfloor$
4:    Let set $K$ be the indices of the $n$ largest $f_j$
5:    **if** $j \in K$ **then**
6:       $\hat{x}_{ij} \leftarrow \lceil x^*_{ij} \rceil$
7:    **else**
8:       $\hat{x}_{ij} \leftarrow \lfloor x^*_{ij} \rfloor$
9:    **end if**
10: **end for**

---

    Algorithm 1 operates on each row of $x^*$ independently. The variable $n$ is the number of assignments in a row that must be rounded up to satisfy the task assignment constraint. Let $f_j$ be the fractional part of the number of tasks that must be assigned to machine $j$. The algorithm simply rounds up those $n$ assignments that have the largest fractional parts. Everything else is rounded down. The result is an integer solution $\hat{x}$ that still assigns all tasks properly and is near to the original solution from the lower bound. Algorithm 1 minimizes the $L_1$ norm between the integer solution and the real-valued solution because the $L_1$ norm is separable and this algorithm chooses the $K$ dimensions to round up that will introduce the least error per dimension.

To illustrate the behavior of the algorithm, let the input be given by Eq. (5). The values in bold indicate assignments that are to be rounded up. The output of the algorithm is given in Eq. (6). The first row does not need to be rounded. The second row rounds up 9.6 because $0.6 \geq 0.4$ and rounds every other component down. The third row shows that the algorithm is not traditional rounding because it rounds up 11.4 due to $0.4 \geq 0.3$. The last row shows how the algorithm might round up two values ($n = 2$).

$$x^* = \begin{pmatrix} 3 & 0 & 9 & 11 & 0 & 0 \\ 3 & 0 & \mathbf{9.6} & 11.4 & 0 & 0 \\ 3 & 15.3 & 9.3 & \mathbf{11.4} & 0 & 0 \\ 3 & 15.2 & \mathbf{9.9} & \mathbf{11.4} & 2.3 & 4.2 \end{pmatrix} \tag{5}$$

$$\hat{x} = \begin{pmatrix} 3 & 0 & 9 & 11 & 0 & 0 \\ 3 & 0 & 10 & 11 & 0 & 0 \\ 3 & 15 & 9 & 12 & 0 & 0 \\ 3 & 15 & 10 & 12 & 2 & 4 \end{pmatrix} \tag{6}$$

The makespan computed from the integer solutions produced by Algorithm 1 may still not be realizable, even though an integer number of tasks is assigned to machine types. To obtain the makespan of the integer solution, computed by Eq. (2), one might still be forced to split tasks among machines of a given machine type to force the finishing times of all the machines to be the same. The local assignment algorithm, discussed in the next subsection, will remedy this by forcing each task to be wholly assigned to a single machine.

## 2.5 Local Assignment

The last phase in recovering a feasible assignment solution is to schedule the tasks already assigned to each machine type to specific machines within that group of machines. This scheduling problem is much easier than the general case because the execution and energy characteristics of all machines in a group are the same. This problem is formally known as the multiprocessor scheduling problem [10]. One must schedule a set of heterogeneous tasks on a set of identical machines. The *longest processing time* (LPT) algorithm is a very common algorithm for solving the multiprocessor scheduling problem [10]. Algorithm 2 uses the LPT algorithm to independently schedule each machine type.

---

**Algorithm 2** Assign tasks to machines using LPT algorithm for each machine type

---

1: **for** $j = 1$ to $M$ **do**
2:   Let $z$ be an empty list
3:   **for** $i = 1$ to $T$ **do**
4:     $z \leftarrow$ join($z$, (task type $i$ replicated $\hat{x}_{ij}$ times))
5:   **end for**
6:   $y \leftarrow$ sort$_{\text{descending by ETC}}(z)$
7:   **for** $k = 1$ to $\| y \|$ **do**
8:     Assign task $y_k$ to the earliest ready time machine of type $j$
9:     Update ready time
10:   **end for**
11: **end for**

---

Each column of $\hat{x}$ is processed independently. List $z$ contains $\hat{x}_{ij}$ tasks for each task type $i$. The tasks are then sorted in descending order by execution time. Next the algorithm loops over this sorted list one element (task) at a time and assigns it to the machine that has the earliest ready time. The *ready time* of a machine is the time at which all tasks assigned to it will complete. This heuristic packs the largest tasks first in a greedy manner. Algorithms exist that will produce a more optimal solution, but it will be shown that the effect of the sub-optimality of this algorithm on the overall performance of the systems diminishes as the problem sizes become large.

## 2.6 Complexity Analysis

The complexity analysis of this algorithm shows some desirable properties that are now discussed. One must solve a real-valued LP problem to compute the lower bound. Using the simplex algorithm to solve the LP problem yields exponential complexity (i.e., traversing all the vertices of the polytope) in the worst case; however the average case complexity for a very large class of problems is polynomial time. Recall that there are $T$ task types and $M$ machine types. The lower bound LP problem has $T + M$ nontrivial constraints and $TM + 1$ variables. The average case complexity of computing the lower bound is $(T + M)^2(TM + 1)$. Next is the rounding algorithm. The outer loop iterates $T$ times, and the rounding is dominated by the sorting of $M$ items. Thus the complexity of Algorithm 1 is $T(M \log M)$. The task assignment algorithm outer loop is run $M$ times. Inside this loop there are two steps. The first step is sorting $n_j = \sum_i x_{ij}$ items which takes $n_j \log n_j$ time. The second step is a loop that iterates $n_j$ times and must find the machine with the earliest ready time each iteration, which is a $\log M_j$ time operation. The worst case complexity of Algorithm 2 is thus $M \max_j \left( n_j \log n_j + n_j \log M_j \right)$.

The complexity of the overall algorithm to find both the lower bound and upper bound (full allocation) is driven by either the lower bound algorithm or the local assignment algorithm. Complexity of the lower bound and Algorithm 1 are independent of the number of tasks and machines. Those algorithms depend only on the number of task types and machine types. This is a very important property for large scale systems. Millions of tasks and machines can be handled easily so long as the machines can be reasonably placed in a small number of homogeneous groups and, likewise, tasks can be grouped by a small number of task types. Only the upper bound's complexity has a dependence on the number of tasks and machines. This phase is only necessary if a full allocation or schedule is required. Furthermore, Algorithm 2 can be trivially parallelized because each machine group is scheduled independently. The lower bound can be used to analyze much of the behavior of the system at less computational cost.

## 3 Scaling Results

An important property of a scheduling algorithm is its ability to scale well as the size of the problem grows. Simulation experiments were carried out to quantify how the relative error and the computational cost of the algorithm scales. These experiments are used to validate the complexity analysis results from Sect. 2.6. $ETC$ and $APC$ are needed to test the algorithms. A set of five benchmarks executed over nine machine types where used to construct the initial matrices [11]. Then the method found in [7] was used to construct larger $ETC$ and $APC$ matrices. Nominally there are 1,100 tasks made up of 30 task types. The number of tasks per task type varies from 11 to 75. There are nine machine types with four machines of each type for a total of 36 machines. A complete description of the systems and output from the algorithms are available in [12].

The number of tasks, task types, and machine types are swept independently to generate a family of figures. For this size system it is intractable to solve for the optimal makespan. It is even too expensive to solve the linear programming relaxation of the assignment of individual tasks to individual machines for this system. This highlights the need for much more scalable solutions. Even though the optimal solution is not known it is still possible to compare bounds on the makespan to gain insight into the algorithm. Each of the three parameter sweeps is computed by taking random subsets with replacement to handle the sweep variable. These results are averaged over 50 Monte Carlo trials. The experiments where performed on a mid-2009 MacBook Pro with a 2.5 GHz Intel Core 2 Duo processor. The code is written in Mathematica 9 and the LP solver uses the simplex method which forwards to the C++ COIN-OR CLP solver [13]. The scaling experiments all optimize makespan while ignoring the energy objective.

Figure 1a shows the relative change in makespan as the number of tasks increase. The number of task types, machines, and machine types are held constant and are the same as the original nine machine type system. The relative increase in makespan is shown from the makespan lower bound ($MS_{LB}$) to the makespan after rounding. Also shown is the increase in makespan from the integer solution to the full allocation. The relative increase in makespan from the lower bound to the upper bound or full

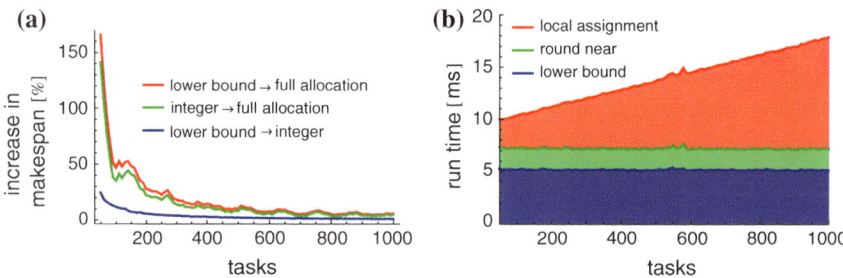

**Fig. 1** Sweeping the *total number of tasks*: **a** shows the relative percent increase in makespan. The quality of the solution improves as more tasks are used. **b** shows the algorithms' run time. Both the lower bound and the rounding algorithms are independent of the number of tasks. The local assignment, used to obtain the full allocation, is linearly dependent on the number of tasks

allocation is also shown. The loss in quality of the makespan from the rounding algorithm is relatively low. Most of the increase in makespan is caused by Algorithm 2. However, Fig. 1a also shows that the relative increase in makespan diminishes as the number of tasks increase. This is because the approximation that tasks are divisible has less of an impact on the solution as the number of tasks increase.

The run time of the scheduler as a function of the number of tasks is shown in Fig. 1b to quantify computational efficiency of the various algorithms. The blue (bottom) portion of the graph is the time taken to compute the lower bound (solve the LP problem). The green (middle) portion is the time it takes to round the solution. Both of the computations required to compute the lower bound and the integer solution do not depend on the number of tasks. This corresponds to the results derived for the complexity of the algorithm. The red (top) portion of the figure shows the full allocation that seems to scale linearly with the number of tasks. Recall that the complexity of Algorithm 2 has a dependency on the number of tasks which is linear or log linear depending on the parameters.

The relative increase in makespan is shown in Fig. 2a. The figure shows the same three curves as Fig. 1a, however this time varying the total number of machines. The number of tasks, machine types, and task types are held constant. As the number of machines increases the increase in makespan due to the full allocation step increases rapidly. This is caused by assigning fewer tasks to each machine as the number of machines increases. The approximation that tasks are divisible becomes a worse approximation as the number of machines increases, relative to the number of tasks.

Figure 2b shows the run time of the three parts of the scheduling algorithm as the total number of machines is varied. Both the lower bound and the rounding are independent of the number of machines. The full allocation step is roughly linear in the number of machines. This corresponds to the analysis in Sect. 2.6.

Figure 3a shows the same three curves as Fig. 1a, however this time varying the number of task types. The number of tasks, machines, and machine types are held constant for this experiment. Figure 3a shows that again the local assignment algorithm is causing most of the degradation in makespan. The relative error in makespan

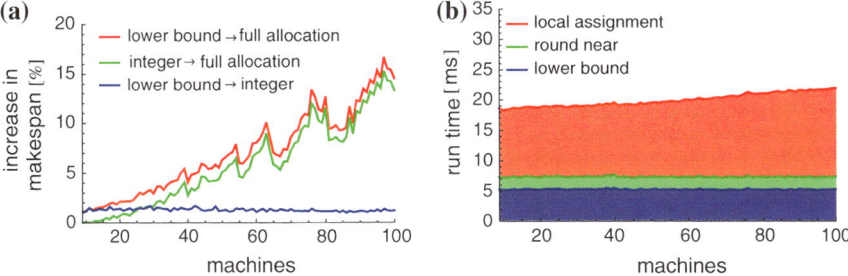

**Fig. 2** Sweep of the *total number of machines*: **a** shows the relative percent increase in makespan. The quality of the solution decreases as more machines are used. **b** shows the algorithms' run time. Both the lower bound and the rounding algorithms are independent of the number of machines. The local assignment, used to obtain the full allocation, is linearly dependent on the number of machines

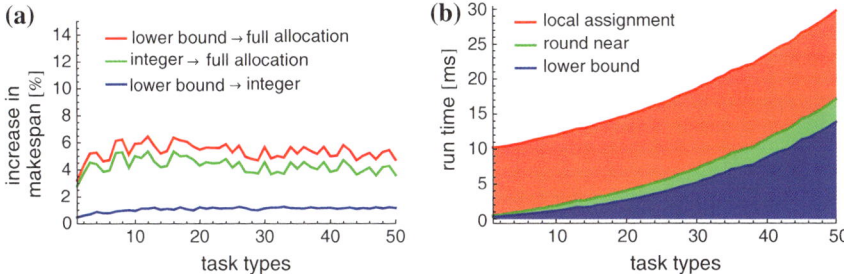

**Fig. 3** Sweeping *number of task types*: **a** shows the relative percent increase in makespan. Quality of the solutions are roughly independent of the number of task types. **b** shows the algorithm's run time. The complexity of the lower bound and rounding algorithms grows super linearly with the number of task types. The local assignment algorithm run time is independent of the number of task types

does not tend to zero because increasing the number of task types does not improve the quality of the approximation.

Figure 3b shows the run time of the three phases. Here the lower bound has super linear dependence on the number of task types. According to the complexity analysis this should be cubic. The rounding algorithm seems to increase linearly, which corresponds to the analysis. The full allocation phase seems to be independent of the number of task types. This agrees with the analysis because the complexity is not a function of the number of task types $T$, but instead a function of the number of tasks $n_j$ assigned to a machine type, regardless of the type of task.

Figure 4a shows the relative increase in makespan as the number of machine types varies. In the previous parameter sweeps, the number of tasks of a particular type may be zero if the random sampling selected that configuration. Allowing the number of machines in a machine type to be zero is troublesome due to Eq. (1) because some constraint coefficients will be $\infty$ in the linear programming problem. Practically, an $M_j = 0$ means that the $j$th column of $\boldsymbol{ETC}$ and $\boldsymbol{APC}$ should simply be removed

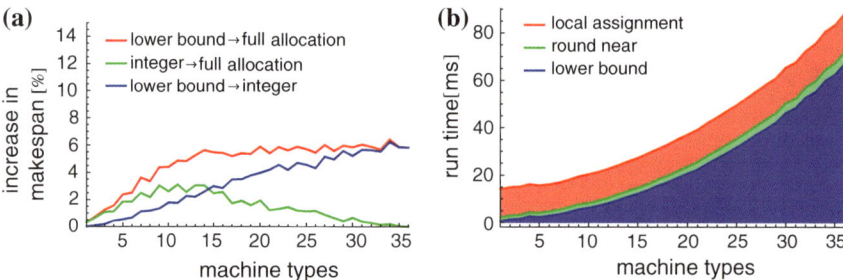

**Fig. 4** Sweeping the *number of machine types*: **a** shows the relative percent increase in makespan. Overall performance is roughly independent of the number of machine types. **b** shows the algorithm's run time. The lower bound algorithm's complexity is super linear in the number of machines types. The rounding and local assignment algorithms are roughly independent of the number of machine types

and the solution should never assign a task to that machine type because it has no machines. To avoid this case altogether each machine type is forced to have at least one machine (so that there are no degenerate machine types). Figure 4a also shows that the quality of the rounding algorithm decreases as the number of machine types increase. This is expected because there are less tasks to assign to each machine's type making the approximation weaker. At 36 machine types there is exactly one machine per machine type. There is only one solution to that scheduling problem (assign all tasks to the one machine), resulting in no increase in makespan in that phase.

Figure 4b shows the run time as the number of machine types is increased. As expected, the lower bound grows roughly cubically. The rounding algorithm grows roughly linearly also as expected. The time spent performing local assignment for each machine type decreases because fewer tasks are scheduled to less machines as the number of machine types increases so it effectively has little dependence on the number of machine types.

Even though the performance of these polynomial time algorithms are desirable, there is some prior work on theoretical bounds that should be noted. In [14] it is proven that there exists no polynomial algorithm that can provably find a schedule that is less than 3/2 the optimal makespan, unless $P = NP$. Even though Figs. 1–4 suggest that one can do better then 3/2, this is only the case on average. In the next section these algorithms are used to generate Pareto fronts.

## 4 Pareto Front Generation

Multi-objective optimization is challenging because there is usually no single solution that is superior to all others. Instead, there is a set of superior feasible solutions that are referred to as the non-dominated solutions [15]. Feasible solutions that are dominated are generally of little interest because one can always find a better solution in some or all objectives by selecting a solution from the non-dominated set. When all objectives should be minimized, a feasible solution $x$ dominates a feasible solution $y$ when:

$$\forall i \quad f_i(x) \le f_i(y)$$
$$\exists i \quad f_i(x) < f_i(y) \tag{7}$$

where $f_i(\cdot)$ is the $i$th objective function. The non-dominated solutions, also known as *outcomes*, compose the Pareto front.

Finding the Pareto front can be computationally expensive because it involves solving numerous variations of the optimization problem. Most algorithms use scalarization techniques to convert the multi-objective problem into a set of scalar optimization problems. Major approaches of scalarization include the hybrid method [16], elastic constraint method [16], Benson's algorithm [17, 18], and Pascoletti-Serafini scalarization [19]. Pascoletti-Serafini scalarization is a generalization of many common approaches such as normal boundary intersection, $\epsilon$-constraint, and weighted sum. We will use the weighted sum algorithm in this work. The weighted sum algorithm can find all the non-dominated solutions for problems with a convex constraint set and convex objective functions [19]. Weighted sum is used for the linear convex problem in Eq. (4) to find all non-dominated solutions. A known issue with the weighted sum algorithm is that it does not uniformly distribute the solutions along the Pareto front. The solutions are often clustered together, but this does not present a problem for our particular use case.

Finding the optimal schedule for makespan alone is NP-Hard in general [20], thus finding the optimal (true) Pareto front is NP-Hard as well. However, computing tight upper and lower bounds on the Pareto front is still possible. Specifically, a lower bound on a Pareto front is a set of solutions for which no feasible solution dominates any of the solutions in this set. An upper bound on the Pareto front is a set of feasible solutions that do not dominate any Pareto optimal solutions. The true Pareto front only exists between the lower bound curve and the upper bound curve.

## 4.1 Weighted Sum

The weighted sum algorithm simply forms the convex combination of the objectives and sweeps the weights to generate the Pareto front. The first step is to compute the lower bound solution for energy and makespan independently of each other. Next let $\Delta E_{LB}$ be the difference in energy between these two lower bound solutions. Likewise, let $\Delta MS_{LB}$ be the difference in makespan between these two lower bound solutions. The scalarized objective is:

$$\min \frac{\alpha}{\Delta E_{LB}} E_{LB} + \frac{1-\alpha}{\Delta MS_{LB}} MS_{LB}. \tag{8}$$

A lower bound on the Pareto front can be generated by using several values of $\alpha \in [0, 1]$. Weighted sums will produce duplicate solutions (i.e., $x$ is identical for neighboring values of $\alpha$). Duplicate solutions are removed to increase the efficiency of the subsequent algorithms. Each solution is rounded by Algorithm 1 to generate an intermediate Pareto front. Rounding often introduces many duplicates that can be safely removed. Each integer solution is converted to a full allocation with Algorithm 2 to create the upper bound on the Pareto front.

## 4.2 Non-dominated Sorting Genetic Algorithm II

NSGA-II [21] is an adaptation of the Genetic Algorithm (GA) optimized to find the Pareto front of a multi-objective optimization problem. Similar to all GAs, the NSGA-II uses mutation and crossover operations to evolve a population of chromosomes (solutions). Ideally this population improves from one generation to the next. Chromosomes with a low fitness are removed from the population. The NSGA-II algorithm modifies the fitness function to work well for discovering the Pareto front. In prior work [3], the mutation and crossover operations were defined for this problem. The NSGA-II algorithm will be seeded in two ways in the following results. The first seeding method is to use the optimal minimum energy solution, sub-optimal minimum makespan solution (from the Min-Min Completion Time [4] algorithm), and a random population as the initial population. This is the original seeding method used in [3]. The second seeding method is to use the full allocations from Algorithm 2 as the initial population for the NSGA-II.

## 5 Pareto Front Results

The system used for these experiments is the same as in Sect. 3, unless stated otherwise. All 1,100 tasks, 30 task types, 36 machines, and nine machines types are used and are described in [8]; the complete description of the system and output data files from the new algorithm are available in [12]. The hardware used for running the NSGA-II experiments is a 2013 Dell XPS' 15 with an Intel i7-4702HQ 2.2 GHz CPU. The NSGA-II code is implemented in C++.

Figure 5 shows Pareto fronts for four different systems. Pareto fronts for each of the algorithms are shown for each system. All the systems have zero idle power consumption.

Figure 5a shows the Pareto fronts generated from the lower bound, round near, local assignment, and NSGA-II algorithms. The figure shows the actual solutions as markers that are connected by lines. The legend shows the total time duration

**Fig. 5** Pareto front for lower bound, integer, upper bound, and NSGA-II: the lower bound does truly lower bound the other curves. The full allocation or upper bound is very near the lower bound so the optimal Pareto front is tightly bounded. The times shown in the parenthesis in the legend indicate the total time to compute the solution. The NSGA-II with the original seed solution quality is rather poor and expensive to compute, however the NSGA-II seeded with the full allocations produces a reasonable result, close to the full allocation, in much less time, but still not as good as the full allocation in places **a** nine machine type system, **b** six machine type system, **c** two machine type system, **d** synthetic ten machine type system

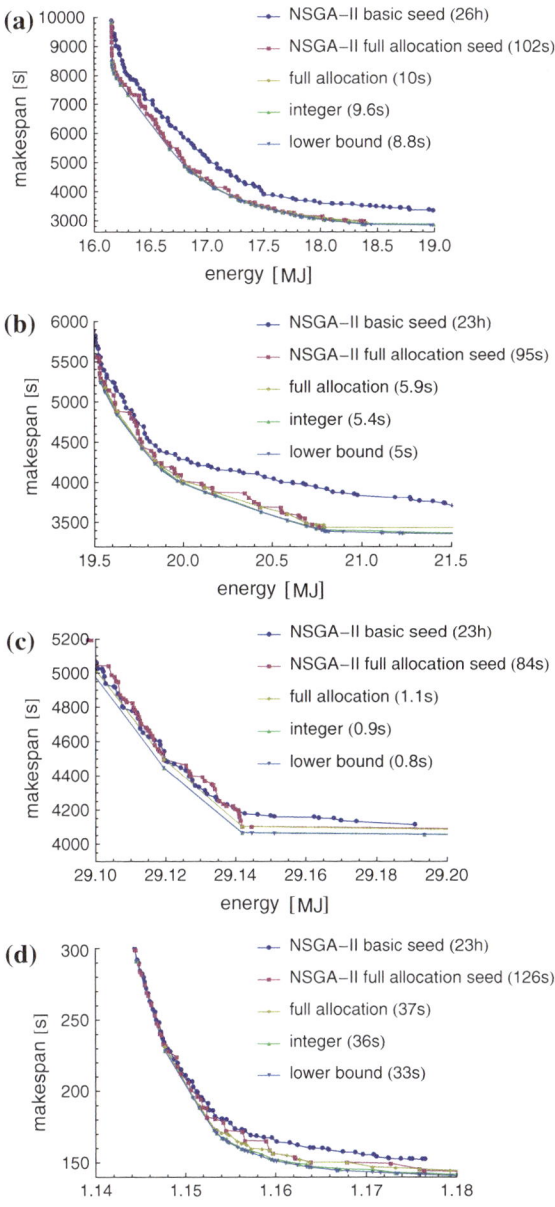

for computing a given Pareto front. For instance having already computed the lower bound the additional time necessary to compute the full allocation is 1.2 s. The lower bound, integer, and full allocation are nearly indistinguishable along the entire Pareto front. This means that the true Pareto front is tightly bounded even though it is unknown. The curve that is dominated by all other curves is the Pareto front generated by the NSGA-II using the first seeding method. The NSGA-II took hours to find that sub-optimal Pareto front. In contrast, the lower and upper bounds were found in approximately 10 s. The last Pareto front is the NSGA-II seeded with the full allocation. Seeding with the full allocation allows the NSGA-II to both converge to an improved Pareto front as well as decrease the run time necessary to converge. The NSGA-II attempts to evenly distribute the solutions along the Pareto front as can be seen in Fig. 5a. All the algorithms seem to perform well when minimizing energy alone because computing the optimal minimum energy solution is relatively easy. One simply assigns each task to the machine that requires the lowest energy to execute that task. Solving for the optimal makespan is difficult in practice. Figure 5a shows that all the algorithms agree in the energy dimension, however in the makespan dimension there are significant distinctions in solution quality. The new algorithms produce better quality Pareto fronts in significantly less time.

A few different systems are used to further illustrate the applicability of the LP-based Pareto front generation technique. Figure 5b shows a system composed of the first six machine types from the previous system, with six machines per type. Figure 5c shows an even smaller system by taking only the first two machine types, with 18 machines per type. The total number of tasks, task types, and machines is unchanged. These figures show how the lower bound and upper bound can still outperform the NSGA-II algorithm even when the number of machines types become small.

The results in Fig. 5d are based on an entirely different system that was previously used in [3]. The system has 50 machines composed from ten machine types. There are 1,000 tasks made from 50 task types. The ETC and APC matrices were generated randomly with the Coefficient of Variation (COV) method described in [22]. Even though this system is very different from the previous systems, the LP-based algorithm produces a superior Pareto front in significantly less time.

Figure 6a illustrates how the solutions progress through the three phases of the algorithm when there is no idle power consumption. The lowest line represents the lower bound on the Pareto front. Each orange arrow represents a solution as it is rounded. In every case, the makespan increases but the energy may increase or decrease. As $x$ is rounded, machines will finish at different times increasing the makespan. Each blue arrow represents a solution that is being fully allocated via the local assignment algorithm. The energy in this case does not change because the local assignment algorithm does not move tasks across machine types thus the power consumption cannot change. The makespan increases are highly varying. The full allocation solution second to the right dominates the one on the far right. In this case the solution on the far right is taken out of the estimate of the Pareto front.

Figure 6b shows the progression of the the the solutions with non-zero idle power. The idle power consumption is set to 10 % of the mean power for each machine type, specifically $APC_{\emptyset j} = \frac{0.1}{T} \sum_i APC_{ij}$. As the makespan increases, more machines will be idle for longer, so the idle energy increases. The local assignment phase now negatively affects the energy consumption because it will may have machines idle for some amount time.

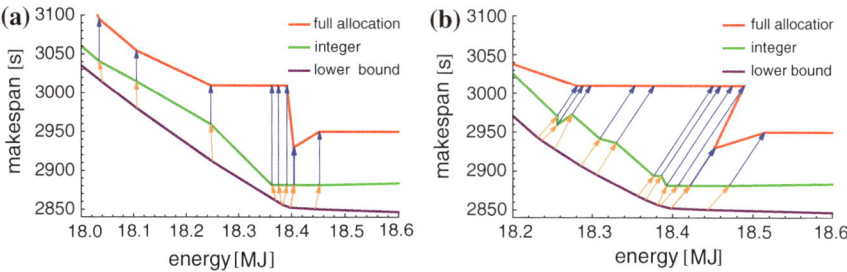

**Fig. 6** Progression of solutions from lower bound to integer to upper bound **a** no idle power **b** 10 % idle power

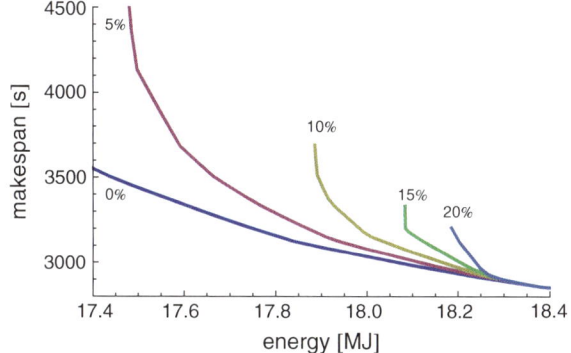

**Fig. 7** Pareto front lower bounds when sweeping idle power: Idle power is swept from 5 % increments as labeled on the figure. As idle power increases the reward for improving makespan also increases

Figure 7 shows the effect of idle power on the Pareto front. The curves show the lower bound on the optimal Pareto front with different idle powers. The penalty for having a large makespan increases as the idle power increases. The optimal energy solutions now must have a shorter makespan to reduce energy usage. This causes the Pareto front to contract in the makespan dimension and shift to the right slightly.

## 6 Related Work

Techniques for generating Pareto fronts have been well studied [3, 7, 8, 15, 21]. This work achieves huge gains in performance over prior algorithms by exploiting properties specific to this static scheduling problem.

Our approach takes advantage of the common property that each machine in an HPC system is not unique but belongs to one of a few machine types. Our work also is focused on very large-scale systems and how to find high quality solutions on average, whereas [20, 23] are concerned with worst-case performance of the scheduling algorithms. The energy and makespan problem is a specialization of the classic optimization problem of minimizing makespan and cost [20, 23].

While this paper deals with scheduling tasks to entire machines, the algorithms in this paper could also be applied to scheduling tasks to cores within a machine or across cores on many machines. The full allocation recovery algorithm we use is similar to the algorithms presented in [24] that deal with scheduling on a single machine by using dynamic voltage and frequency scaling (DVFS).

# 7 Conclusion

A highly scalable scheduling algorithm for the energy and makespan bi-objective optimization problem was presented. The complexity of the algorithm to compute the lower bound on the Pareto front was shown to be independent of the number of tasks. The quality of the solution also improves as the size of the problem increases. These two properties make this algorithm perfectly suited for very large scale scheduling problems. Algorithms were also presented that allow one to efficiently recover feasible solutions. These feasible solutions serve as the upper bound on the Pareto front and can be used to seed other algorithms. This upper bound was compared to the solution found with the NSGA-II algorithm and shown to be superior in solution quality and algorithm run time. These algorithms allow the decision makers to more easily trade-off energy and makespan to reduce operating costs and improve efficiency of HPC systems.

This work could be extended by considering alternative scalarization techniques to hopefully reduce the time required to compute the lower bound. Many of the LP problems result in solutions that are identical thus providing minimal information in forming the Pareto front. It is possible to avoid generating duplicate solutions by utilizing different scalarization techniques. The LP-based scheduling algorithm only takes a fraction of a second to compute a single schedule for a given bag-of-tasks so it is possible to use this scheduler for online batch-mode scheduling. Specifically this algorithm can be used to schedule tasks as they arrive at the system by computing a schedule for all tasks waiting in the queue (as a batch) and recomputing the schedule when a task completes or a new task arrives.

**Acknowledgments** This work was supported by the Sjostrom Family Scholarship, Numerica Corporation, the National Science Foundation (NSF) under grants CNS-0905399 and CCF-1302693, the NSF Graduate Research Fellowship, and by the Colorado State University George T. Abell Endowment. Any opinion, findings, and conclusions or recommendations expressed in this material are those of the authors and do not necessarily reflect the views of the National Science Foundation. A preliminary version of portions of this work have been previously presented in [25].

# References

1. Koomey, J.: Growth in data center electricity use 2005 to 2010, pp. 1. Analytics Press (2011)
2. Cameron, K.W.: Energy oddities, part 2: why green computing is odd. Computer **46**(3), 90–93 (2013)
3. Friese, R., Brinks, T., Oliver, C., Siegel, H.J., Maciejewski, A.A.: Analyzing the trade-offs between minimizing makespan and minimizing energy consumption in a heterogeneous resource allocation problem. In: INFOCOMP, The Second International Conference on Advanced Communications and Computation. 81–89 (2012)
4. Braun, T.D., Siegel, H.J., Beck, N., Bölöni, L.L., Maheswaran, M., Reuther, A.I., Robertson, J.P., Theys, M.D., Yao, B., Hensgen, D., Freund, R.F.: A comparison of eleven static heuristics for mapping a class of independent tasks onto heterogeneous distributed computing systems. J. Parallel Distrib. Comput. **61**(6), 810–837 (2001)
5. Bharadwaj, V., Robertazzi, T.G., Ghose, D.: Scheduling Divisible Loads in Parallel and Distributed Systems. IEEE Computer Society Press, Los Alamitos (1996)
6. Al-Qawasmeh, A.M., Maciejewski, A.A., Wang, H., Smith, J., Siegel, H.J., Potter, J.: Statistical measures for quantifying task and machine heterogeneities. J. Supercomput. **57**(1), 34–50 (2011)
7. Friese, R., Khemka, B., Maciejewski, A.A., Siegel, H.J., Koenig, G.A., Powers, S., Hilton, M., Rambharos, J., Okonski, G., Poole, S.W.: An analysis framework for investigating the trade-offs between system performance and energy consumption in a heterogeneous computing environment. In: IEEE 27th International Parallel and Distributed Processing Symposium Workshops (IPDPSW), Heterogeneity in Computing Workshop, IEEE, pp. 19–30 (2013)
8. Friese, R., Brinks, T., Oliver, C., Siegel, H.J., Maciejewski, A.A., Pasricha, S.: A machine-by-machine analysis of a bi-objective resource allocation problem. In: International Conference on Parallel and Distributed Processing Technologies and Applications (PDPTA) (2013)
9. Bertsimas, D., Tsitsiklis, J.N.: Introduction to Linear Optimization. Optimization and Neural Computation. Athena Scientific (1997)
10. Graham, R.: Bounds on multiprocessing timing anomalies. SIAM J. Appl. Math. **17**(2), 416–429 (1969)
11. Phoronix Media: Intel core i7 3770k power consumption, thermal. http://openbenchmarking.org/result/1204229-SU-CPUMONITO81 (May 2013)
12. Tarplee, K.M.: Energy and makespan bi-objective optimization data. http://goo.gl/3Ik8eC (November 2013)
13. Hall, J.: Coin-or clp. https://projects.coin-or.org/Clp (March 2013)
14. Lenstra, J., Shmoys, D., Tardos, É.: Approximation algorithms for scheduling unrelated parallel machines. Math. Program. **46**(1–3), 259–271 (1990)
15. Pareto, V.: Cours d'economie Politique. F. Rouge, Lausanne (1896)
16. Ehrgott, M.: Multicriteria Optimization. Springer-Verlag New York Inc, Secaucus (2005)
17. Benson, H.: An outer approximation algorithm for generating all efficient extreme points in the outcome set of a multiple objective linear programming problem. J. Global Optim. **13**(1), 1–24 (1998)
18. Löhne, A.: Vector Optimization with Infimum and Supremum. Vector Optimization. Springer, Berlin Heidelberg (2011)
19. Eichfelder, G.: Adaptive Scalarization Methods in Multiobjective Optimization. Springer (2008)
20. Jansen, K., Porkolab, L.: Improved approximation schemes for scheduling unrelated parallel machines. Math. Oper. Res. **26**(2), 324–338 (2001)
21. Deb, K., Pratap, A., Agarwal, S., Meyarivan, T.: A fast and elitist multiobjective genetic algorithm: NSGA-II. IEEE Trans. Evol. Comput. **6**(2), 182–197 (2002)
22. Ali, S., Siegel, H.J., Maheswaran, M., Hensgen, D., Ali, S.: Representing task and machine heterogeneities for heterogeneous computing systems. Tamkang J. Sci. Eng. **3**(3), 195–208 (2000)

23. Shmoys, D.B., Tardos, É.: Scheduling unrelated machines with costs. In: Fourth Annual ACM-SIAM Symposium on Discrete algorithms, Society for Industrial and Applied Mathematics. 448–454 (1993)
24. Li, D., Wu, J.: Energy-aware scheduling for frame-based tasks on heterogeneous multiprocessor platforms. In: 41st International Conference on Parallel Processing (ICPP), pp. 430–439 (2012)
25. Tarplee, K.M., Friese, R., Maciejewski, A.A., Siegel, H.J.: Efficient and scalable computation of the energy and makespan pareto front for heterogeneous computing systems. In: Federated Conference on Computer Science and Information Systems, Workshop on Computational Optimization. 401–408 (2013)

# Measuring Performance of a Hybrid Optimization Algorithm on a Set of Benchmark Functions

**Ezgi Deniz Ulker and Ali Haydar**

**Abstract**  Hybrid algorithms are effective in solving complex optimization problems instead of using traditional methods. In literature, many proposed hybrid algorithms can be seen in order to increase their performance by the use of features of well-known algorithms. The aim of hybridization is to have better solution quality and robustness than traditional optimization algorithms by balancing the exploration and exploitation goals. This paper investigates the performance of a novel hybrid algorithm composed of Differential Evolution algorithm, Particle Swarm Optimization algorithm and Harmony Search algorithm which is called HDPH. This is done on a set of known benchmark functions. The experimental results show that HDPH has a good solution quality and high robustness on many benchmark functions. Also, in HDPH all control parameters are randomized in given intervals to avoid selecting all possible combination of control parameters in given ranges.

**Keywords**  Differential evolution · Evolutionary algorithms · Harmony search · Hybrid optimization algorithms · Particle swarm optimization

## 1 Introduction

In recent years, optimization has become an attractive field for researchers to solve complex functions in different areas [1–4]. Many well-known optimization algorithms are recommended to obtain the optimum value of the problems [5–10]. Some of these algorithms are Genetic Algorithm (GA), Particle Swarm Optimization (PSO), Ant Colony Optimization (ACO), Differential Evolution (DE) and Harmony Search (HS) algorithm.

E.D. Ulker · A. Haydar (✉)
Department of Computer Engineering, Girne American University,
Karaoglanoglu, Mersin 10, Turkey
e-mail: ahaydar@gau.edu.tr

E.D. Ulker
e-mail: ezgideniz@gau.edu.tr

© Springer International Publishing Switzerland 2015
S. Fidanova (ed.), *Recent Advances in Computational Optimization*,
Studies in Computational Intelligence 580, DOI 10.1007/978-3-319-12631-9_11

One of the goals of an optimization algorithm is to keep a balance between exploration and exploitation. The success of the algorithm directly depends on the use of exploration and exploitation features effectively. Exploration is the ability of the optimization algorithm to test effectively the search space for various candidate solutions. The feature of exploitation is the ability of an optimization algorithm to make the search around possible candidate solutions. The optimization algorithms can solve the complex problems by keeping the balance between exploitation and exploration attributes and also by using of their characteristic features. Altough the well known optimization algorithms have positive properties, it is observed that these algorithms do not always perform well on some problems [11]. Because of this, modified algorithms and hybrid algorithms are growing area of interest by researchers. In modified optimization algorithms, only minor changes are applied to characteristics of the original method [12–15], but the main steps of the algorithms are kept intact. In contrast, in hybrid optimization algorithms, two or more different techniques are selected and combined without altering their main characteristic features.

There are many ways to generate a hybrid optimization algorithm known in literature, and it is observed that hybridization is quite effective for optimization[15–19]. In this study, the performance of a new hybrid algorithm is measured with the powerful optimization algorithms used to form it. The proposed algorithm is called Hybrid Differential Particle Harmony (HDPH) algorithm. It is a combination of three well known evolutionary algorithms, Differential Evolution (DE) algorithm, Particle Swarm Optimization (PSO) algorithm, and Harmony Search (HS) algorithm. In the HDPH algorithm, properties of the three algorithms that form it are not altered and applied to candidates recursively. This helps in the achievement of desired exploration and exploration features of HDPH algorithm.

The rest of the paper is organized as follows. Section 2 describes the HDPH algorithm and its main operators in detail. Section 3 describes measurement of performance of the HDPH algorithm and compares it with performances of the three algorithms that form it with the discussion of the obtained results. Section 4 concludes the paper.

## 2 The HDPH Algorithm

In literature, various optimization algorithms are suggested to generate hybrid algorithms for more efficient optimization [15–19]. The main goal of hybridization is to improve solution quality by combining different features of several algorithms. Such well-known and powerful algorithms as DE, PSO and HS are used in many different fields by researchers and proved to be quite effective in solving various types of problems [6, 7, 14]. Details of algorithms DE, PSO and HS are given in publications [1–3], respectively. Here, we briefly explain how these algorithms are used in our hybrid algorithm HDPH. Each of the mentioned algorithms has different characteristics. The algorithm DE has high solution quality for most of the problems,

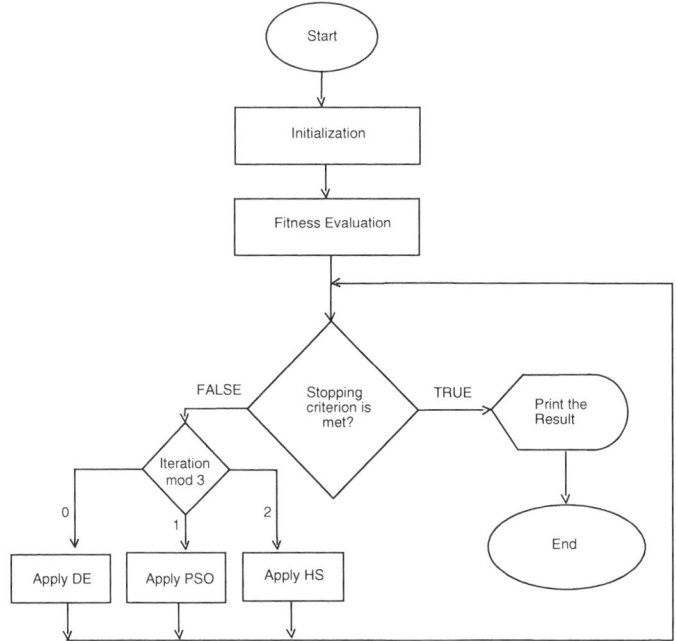

**Fig. 1** Flowchart of the HDPH algorithm

PSO uses its exploration to avoid the solution from trapping into local optimum and HS has good performance and robustness on different problems [20].

HDPH uses main operators of the individual algorithms DE, PSO and HS with randomly selected parameters without modification or change of the main character-istics of these algorithms. The following Fig. 1 shows flowchart of HDPH algorithm. The algorithm can be described with the following steps.

Step 1. *Generation of the candidate population with given dimensions:*
Initialize the candidate population $X_i$ where $i = \{1, 2, 3, \ldots, N\}$ within given ranges and $N$ is the size of the population.

Step 2. *Crossover and mutation operations of DE:*
The mutation and crossover operators are applied by DE to find better approximation to a solution by using the following expressions

$$V_i = X_a + F\,(X_b - X_c),\tag{1}$$

$$U_{ij} = \begin{cases} V_{ij} & \text{if } r_j \le CR, \\ X_{ij} & \text{otherwise,} \end{cases}\tag{2}$$

$$X_i = \begin{cases} U_i & \text{if } f(U_i) < f(X_i), \\ X_i & \text{otherwise.} \end{cases}\tag{3}$$

The mutant vector $V_i$ is calculated using (1) where $X_a$, $X_b$, and $X_c$ are distinct members in the population. In the crossover operator, $V_{ij}$ is crossovered with $X_{ij}$ to generate $U_{ij}$ by using (2), where $V_{ij}$ and $X_{ij}$ are the $j_{th}$ elements of the $i_{th}$ mutant vector $V_i$ and $i_{th}$ solution vector $X_i$, respectively and $r_j$ is uniformly distributed number for each $j_{th}$ element of $V_i$. Here, $F$ and $CR$ are control parameters of DE. They are used for mutation and crossover operations. In the selection process, new candidates for $X_i$ are determined as either the vector $U_i$ or its previous solution depending on the fitness values of $U_i$ and $X_i$ by using (3).

Step 3. *Computing particle movement by PSO:*
Apply operators of PSO algorithm by using the following expressions

$$V_i = w V_i + c_1 \left( P_{\text{best},i} - X_i \right) + c_2 \left( \text{global}_{\text{best}} - X_i \right), \tag{4}$$

$$X_i = X_i + V_i. \tag{5}$$

Particles improve their positions by using (5). This movement excludes the particles to be trapped to the local optimum by increasing the diversity of solution vector. $V_i$ refers to the velocity value of particle $i$ for each row and it is calculated according to the control parameters $c_1$, $c_2$, and $w$ by using (4). $\text{global}_{\text{best}}$ is the best known position reached up to that point and $P_{\text{best},i}$ is the best position reached by the $i_{th}$ particle in the swarm. $X_i$ refers to the position of particle $i$ and can be updated by using (5) for each row.

Step 4. *Choosing a neighboring value by HS:*
HS can search in different zones of the search space by using the control parameters that are *hmcr*, *par* and *fw*. With probability of *hmcr* a candidate member is selected from the population. With probability of 1-*hmcr* a candidate is generated randomly and added into the population. HS can have non-updated candidate elements in the population with probability of 1-*par*. The $j_{th}$ element of the candidate $i$, $X_i$ can be updated with probability of *par* by applying the following expression

$$X_{ij} = X_{ij} + rand \; () \; fw, \tag{6}$$

where *rand*() is a random number in the range $\in (-1, 1)$.

Step 5. Repeat Steps 2, 3 and 4 until the chosen stopping criterion is met.
The algorithm is performed in a loop until the termination criterion is satisfied. Elitism is included in HDPH by keeping the best solution at the end of each iteration.

## 3 The Numerical Results and Discussions

The proposed hybrid algorithm is tested with the use of 22 well studied benchmark functions with different characteristics. The performances are measured both for HDPH and the 3 algorithms that form it in terms of solution quality and robustness for random initialization of the population. The benchmark functions are selected either

**Table 1** Unimodal benchmark functions

| D | Function | Expression | $f_{min}$ |
|---|---|---|---|
| 30 | Step | $\sum_{i=1}^{n} (\lfloor x_i + 0.5 \rfloor)^2$ | 0 |
| 30 | Sphere | $\sum_{i=1}^{n} (x_i^2)$ | 0 |
| 30 | SumSquares | $\sum_{i=1}^{n} (ix_i^2)$ | 0 |
| 30 | Quartic | $\sum_{i=1}^{n} (ix_i^4 + rand\,[0, 1))$ | 0 |
| 6 | Trid6 | $\sum_{i=1}^{n} (x_i - 1)^2 - \sum_{i=2}^{n} (x_i x_{i-1})$ | −50 |
| 10 | Trid10 | $\sum_{i=1}^{n} (x_i - 1)^2 - \sum_{i=2}^{n} (x_i x_{i-1})$ | −210 |
| 10 | Zakharov | $\sum_{i=1}^{n} (x_i^2) + \left(\sum_{i=1}^{n} 0.5ix_i\right)^2 + \left(\sum_{i=1}^{n} 0.5ix_i\right)^4$ | 0 |
| 24 | Powell | $\sum_{i=1}^{n/k} (x_{4i-3} + 10x_{4i-2})^2 +$ <br> $5 (x_{4i-1} - x_{4i})^2 +$ <br> $(x_{4i-2} - x_{4i-1})^4 +$ <br> $10 (x_{4i-3} - x_{4i})^4,$ <br> $k = 4.$ | 0 |
| 30 | Schwefel 2.22 | $\sum_{i=1}^{n} |x_i| + \prod_{i=1}^{n} |x_i|$ | 0 |
| 30 | Schwefel 1.2 | $\sum_{i=1}^{n} \left(\sum_{j=1}^{i} x_j\right)^2$ | 0 |
| 30 | Rosenbrock | $\sum_{i=1}^{n} \left[100 \left(x_{i+1} - x_i^2\right)^2 + (x_{i-1})^2\right]$ | 0 |
| 30 | Dixon-Price | $(x_{1-1})^2 + \sum_{i=2}^{n} i \left(2x_i^2 - x_{i-1}\right)^2$ | 0 |

unimodal (U) or multimodal (M). The lists of unimodal and multimodal benchmark functions are given in Tables 1 and 2, respectively where D is dimensionality of the search space. The population size was fixed at 100 for all algorithms.

In each algorithm section, some control parameters are used to affect performance of DE, PSO and HS algorithms. Two control parameters of DE algorithm, $F$ and $CR$ are selected from the following sets: $F \in \{0.3, 0.5, 0.7, 0.8, 0.9, 1.2, 1.4\}$ and $CR \in \{0.1, 0.2, 0.4, 0.6, 0.8, 0.9\}$.

Three control parameters used in PSO algorithm are selected from the following sets: $c_1, c_2 \in \{0.3, 0.6, 0.9, 1.2, 1.5, 1.8\}$ and $w \in \{0.4, 0.5, 0.6, 0.7, 0.8, 0.9\}$. For the HS algorithm, the control parameters $hmcr$ and $par$ are selected as $hmcr \in \{0.7, 0.8, 0.9, 0.93, 0.96, 0.98\}$ and $par \in \{0.01, 0.02, 0.05, 0.1, 0.2\}$. Control parameter $fw$ is adjusted as 0.01, 0.05, 0.1, and 0.2 times the upper bound of each function in HS. The experimental results for DE, PSO and HS algorithms are obtained by selecting each possible combination of control parameters and running each selection 20 times. The selected control parameter ranges are chosen similarly to the ranges commonly used in literature [1–3]. For each benchmark function, control parameter values that provide the closest approximation of the optimum solution were selected and the function was further evaluated around these selected parameters. By doing this, we try to achieve a good parameter tuning. However, in the proposed HDPH algorithm, instead of selecting values of eight control parameters from the sets used for three algorithms, they were selected randomly in ranges given by the minimal and

**Table 2**  Multimodal benchmark functions

| D | Function | Expression | $f_{min}$ |
|---|----------|------------|-----------|
| 2 | Booth | $(x_1 + 2x_2 - 7)^2 +$ $(2x_1 + x_2 - 5)^2$ | 0 |
| 2 | Rastrigin | $\sum_{i=1}^{n} \left( x_i^2 - 10\cos(2\Pi x_i) \right) + 10$ | 0 |
| 30 | Schewefel | $\sum_{i=1}^{n} -x_i \sin\left( \sqrt{|x_i|} \right)$ | $-418.9D$ |
| 10 | Michalewicz10 | $-\sum_{i=1}^{n} \sin(x_i) \left( \sin\left( \frac{ix_i^2}{\Pi} \right) \right)^{2m},$ $m = 10.$ | $-9.66$ |
| 2 | Schaffer | $\dfrac{0.5 + \left( \sin^2\left( \sqrt{x_1^2 + x_2^2} \right) - 0.5 \right)}{\left[ 1 + 0.001\left( x_1^2 + x_2^2 \right) \right)^2 \right]}$ | 0 |
| 2 | SixHumpCamelBack | $4x_1^2 - 2.1x_1^4 + \frac{1}{3}x_1^6 +$ $x_1 x_2 - 4x_2^2 + 4x_2^4$ | $-1.03163$ |
| 2 | Shubert | $\left( \sum_{i=1}^{5} i\cos(i+1)x_1 + \right.$ $\left. i\left( \sum_{i=1}^{5} i\cos((i+1)x_2 + i) \right) \right)$ | $-186.73$ |
| 30 | Griewank | $1/4000 \left( \sum_{i=1}^{n} x_i^2 - \right.$ $\left. \prod_{i=1}^{n} \cos\left( x_i / \sqrt{i} \right) + 1 \right.$ | 0 |
| 30 | Ackley | $-20\exp\left( -0.2\sqrt{\frac{1}{n} \sum_{i=1}^{n} x_i^2} \right) -$ $\exp\left( \frac{1}{n} \sum_{i=1}^{n} \cos(2\Pi x_i) + 20 + e \right.$ | 0 |
| 30 | Penalized | $\Pi/n \left[ 10\sin^2(\Pi y_1) + \right.$ $\left( \sum_{i=1}^{n-1} (y_i - 1)^2 \left[ 1 + 10\sin^2(\Pi y_{i+1}) \right] \right) +$ $\left. (y_n - 1)^2 \right] + \sum_{i=1}^{n} u(x_i, 10, 100, 4)$ $y_i = 1 + \frac{1}{4}(x_i + 1)$ $u(x_i, a, k, m) = \begin{cases} k(x_i - a)^m, & x_i > a, \\ 0, & -a \leq x_i \leq a, \\ k(-x_i - a)^m, & x_i < a. \end{cases}$ | 0 |

maximal values of each parameter. This is done to reduce the number of combinations of parameter values. The results are obtained only by running the hybrid model 20 times. In Tables 3 and 4, the performances of HDPH, DE, PSO and HS algorithms over 10000 function evaluations are shown. For each algorithm, the best value, the average (Avg) and the standard deviations (Stdev) of the 20 runs for the best combinations of values of parameters are shown. In case that there are more than one combination of control parameter values that give the best value, the one that has the average closest to the optimal value and smaller standard deviation is chosen. The results corresponding to unimodal benchmark functions are presented in Table 3. In Step function, the

**Table 3** Results for unimodal benchmark functions

| Function | Values | HDPH | DE | PSO | HS |
|---|---|---|---|---|---|
| | Avg | 3.3 | 78.7 | 16.8 | 13.4 |
| Step | Stdev | 2.9 | 36.0 | 12.0 | 9.7 |
| | Best | 0 | 52 | 1 | 0 |
| | Avg | 0.04 | 78.4 | 2.1 | 10.4 |
| Sphere | Stdev | 0.1 | 35.4 | 4.8 | 8.7 |
| | Best | 0.0002 | 37.0 | 0.05 | 2.2 |
| | Avg | 0.001 | 8.0 | 0.2 | 0.8 |
| SumSquares | Stdev | 0.002 | 3.7 | 0.2 | 0.2 |
| | Best | $2.06e - 5$ | 3.5 | 0.001 | 0.3 |
| | Avg | 0.007 | 0.006 | 0.005 | 0.001 |
| Quartic | Stdev | 0.026 | 0.002 | 0.002 | 0.0006 |
| | Best | 0.0001 | 0.002 | 0.0007 | 0.0005 |
| | Avg | $-49.8$ | $-50$ | $-49.9$ | $-45.89$ |
| Trid6 | Stdev | 0.106 | 0 | 0.002 | 3.016 |
| | Best | $-49.9$ | $-50$ | $-50$ | $-49.9$ |
| | Avg | $-184.2$ | $-207.8$ | $-209.6$ | $-105.0$ |
| Trid10 | Stdev | 32.5 | 5.6 | 0.86 | 67.7 |
| | Best | $-209.9$ | $-209.9$ | $-210$ | $-209.3$ |
| | Avg | $8.39e - 7$ | $2.63e - 5$ | $1.66e - 6$ | $1.87e - 5$ |
| Zakharov | Stdev | $3.57e - 6$ | $1.4e - 5$ | $4.17e - 6$ | $9.46e - 6$ |
| | Best | $1.83e - 3$ | $9.36e - 6$ | $3.63e - 17$ | $2.55e - 6$ |
| | Avg | 0.1 | 1.5 | 0.97 | 0.95 |
| Powell | Stdev | 0.1 | 1.4 | 1.2 | 0.92 |
| | Best | 0.03 | 0.2 | 0.01 | 0.03 |
| | Avg | 0.001 | 4.9 | 14.5 | 0.34 |
| Schwefel 2.22 | Stdev | 0.002 | 0.4 | 33.3 | 0.12 |
| | Best | $9.27e - 5$ | 4.2 | 0.2 | 0.15 |
| | Avg | 1996.6 | 4817.6 | 1689.4 | 28507.6 |
| Schwefel 1.2 | Stdev | 1234.8 | 1317.6 | 848.0 | 5208.5 |
| | Best | 541.2 | 2983.6 | 358.1 | 14559.0 |
| | Avg | 129.2 | 6988.6 | 387.8 | 983.6 |
| Rosenbrock | Stdev | 81.9 | 3393.4 | 498.3 | 503.5 |
| | Best | 22.2 | 2055.4 | 43.0 | 346.2 |
| | Avg | 3.5 | 64.4 | 29.2 | 10.8 |
| Dixon-Price | Stdev | 1.9 | 31.4 | 113.6 | 3.1 |
| | Best | 0.06 | 25.6 | 0.06 | 3.7 |

speed of convergence of HDPH is substantially better than for each of the other three algorithms. For this function, only HDPH and HS algorithms were able to find the minimum value. The standard deviation value shows that the robustness of HDPH algorithm for random initialization of population outperforms the other algorithms. Similar results are obtained also for Sphere and SumSquares functions. For the Quartic function, even though the best value is obtained by the HDPH algorithm, the standard deviation and the average values show that HS algorithm's performance is better than for the others in terms of speed of convergence and robustness. The best values, obtained by all 4 algorithms approach the minimum value of Trid6 function. However, among them, DE performed best for this function. With obtained standard deviations and averages for Trid10 function, it can be deduced that the convergence speed and robustness of PSO algorithm is better than for the other algorithms even though all of them converge to the minimum of the function. For the Zakharov function, all algorithms performed well. On average, the convergence speed of HDPH is slightly better than that of DE, PSO and HS algorithms. The closest point to the minimum of Powell function is reached by PSO algorithm. For function Schwefel 2.22, none of the algorithms, except HDPH, approached very close the minimum value. One can see that the average and the standard deviation values obtained by HDPH algorithm are much better than the respective values obtained by DE, PSO and HS algorithms.

For Schwefel 1.2 function, none of the algorithms reached the minimum value. When we compare the performances of the algorithms for this function, we see that slightly better results are obtained using PSO algorithm. HDPH algorithm has close performance to the PSO algorithm. For the function Rosenbrock, none of the algorithms approached the optimum value in 10000 function evaluations. However, the smaller average and standard deviation values for the HDPH algorithm show that it is more robust and has faster speed of convergence than the other algorithms. Finally, for Dixon-Price function, the best values are obtained by PSO and HDPH algorithms, but the average and standard deviation of HDPH are considerably better than for each of the merged algorithms.

In Table 4, the results obtained for multimodal benchmark functions are shown. The best values for almost all benchmark functions are obtained by HDPH, except Rastrigin function. These values are either better or similar to the best values obtained by DE, PSO and HS algorithms. In particular, the standard deviations and the averages of HDPH for Schwefel and Michalewicz10 functions are considerably better than for the other three algorithms. However, for Rastrigin function, the standard deviation and the average obtained by HS algorithm are better than for HDPH. For Booth function, all three algorithms have better standard deviation values compared to HDPH.

For Griewank, Ackley, and Penalized functions, the best values obtained by HDPH outperform the other three algorithms. For these three functions, when both the average and standard deviation values are taken into consideration, the HDPH gives better results than DE and PSO algorithms. When it is compared with HS algorithm, except Ackley function which gives similar results, HDPH is again better than HS algorithm. For Schaffer, Six Hump Camel Back and Shubert functions, both the best values and standard deviations are comparable for all four algorithms.

It can be seen from the shown results that HDPH generally works as good as or better than the three algorithms that form it in terms of solution quality and robustness.

**Table 4** Results for multimodal benchmark functions

| Function | Values | HDPH | DE | PSO | HS |
|---|---|---|---|---|---|
| | Avg | 0.0001 | $1.37e-25$ | 0 | $3.36e-7$ |
| Booth | Stdev | 0.0007 | $1.8e-25$ | 0 | $5.02e-7$ |
| | Best | 0 | $2.41e-27$ | 0 | $1.27e-8$ |
| | Avg | 36.1 | 137.8 | 46.3 | 18.1 |
| Rastrigin | Stdev | 14.2 | 6.68 | 17.4 | 3.4 |
| | Best | 21.8 | 126.0 | 19.0 | 12.7 |
| | Avg | $-12567.6$ | $-7485.7$ | $-8531.0$ | $-12554.6$ |
| Schwefel | Stdev | 2.5 | 270.6 | 949.2 | 28.8 |
| | Best | $-12569.5$ | $-8128.5$ | $-10353.9$ | $-12566.1$ |
| | Avg | $-9.65918$ | $-9.13$ | $-8.005$ | $-9.611$ |
| Michalewicz10 | Stdev | 0.0025 | 0.11 | 0.93 | 0.05 |
| | Best | $-9.66015$ | $-9.31606$ | $-9.65524$ | $-9.66004$ |
| | Avg | 0.0079 | 0.001 | 0.0025 | 0.0009 |
| Schaffer | Stdev | 0.0037 | 0.0008 | 0.004 | 0.002 |
| | Best | 0 | $7.8e-5$ | 0 | $1.85e-6$ |
| | Avg | $-1.03163$ | $-1.03163$ | $-1.03163$ | $-1.03163$ |
| SixHumpCamelBack | Stdev | 0 | 0 | 0 | $3.66e-06$ |
| | Best | $-1.03163$ | $-1.03163$ | $-1.03163$ | $-1.03163$ |
| | Avg | $-186.722$ | $-185.624$ | $-186.729$ | $-186.727$ |
| Shubert | Stdev | 0.026 | 1.409 | 0.009 | 0.0043 |
| | Best | $-186.731$ | $-186.703$ | $-186.731$ | $-186.731$ |
| | Avg | 0.04 | 1.53 | 0.39 | 1.004 |
| Griewank | Stdev | 0.07 | 0.2 | 0.3 | 0.022 |
| | Best | 0.0002 | 1.29 | 0.053 | 1.004 |
| | Avg | 0.8 | 16.7 | 2.8 | 1.09 |
| Ackley | Stdev | 0.5 | 0.7 | 1.01 | 0.29 |
| | Best | 0.007 | 15.1 | 0.6 | 0.56 |
| | Avg | 0.03 | 5.08 | 4.3 | 0.2 |
| Penalized | Stdev | 0.05 | 2.1 | 2.9 | 0.2 |
| | Best | $3.59e-6$ | 2.7 | 0.3 | 0.04 |

This is achieved by running the HDPH algorithm only 20 times. However, for DE, PSO and HS algorithms, the tabulated results are obtained by running them 20 times for all possible combinations of parameters, finding the parameter set that gives the best performance, making a parameter tuning around those values and using the parameters that provide the best performance.

# 4 Conclusion

A new hybrid algorithm called HDPH is proposed and investigated in a series of numerical experiments. Being a robust algorithm, it provides good solution quality by combining three existing algorithms, DE, PSO and HS. The performances of used algorithms depend on the parameter selection. Therefore, all combinations of parameter values are experimentally tested for a number of benchmark functions for all algorithms.

The tabulated results represent the best values obtained through all possible trials. However, in the HDPH algorithm the parameters are chosen randomly in the given ranges which makes the algorithm easier to implement. Performance of HDPH algorithm is checked both for unimodal and multimodal benchmark functions. The experimental results show that, for almost all multimodal benchmark functions, the convergence speed and the robustness of the proposed hybrid algorithm is either better or similar to DE, PSO and HS algorithms.

However, for the unimodal benchmark functions, the number of function evaluations was not enough for some functions to arrive closer to the minimum value. Nevertheless, when the average and standard deviation values are taken into consideration, in most cases the proposed hybrid algorithm's performance is better than that of the other algorithms. Furthermore, the experimental results show that the solution quality of the proposed algorithm does not change much with different runs.

For almost all used benchmark functions, its standard deviation values that are obtained by running the algorithm 20 times, are smaller than for DE, PSO and HS algorithms. Thus, on the base of obtained experimental results, one can come to the conclusion that the proposed hybrid algorithm is an acceptable candidate for solving real-time optimization problems.

# References

1. Geem, Z.W., Kim, J.H., Loganathan, G.V.: A new heuristic optimization algorithm: harmony search. Simul., Trans. of the Soc. for Model. Simul. Int. 60–68 (2001)
2. Kennedy, J., Eberhart, R.: Particle swarm optimization. In: IEEE International Conference on Neural Networks, pp. 1942–1948. IEEE Press, New Jersey (1995)
3. Storn, R., Price, K.: Differential evolution; a simple and efficient heuristic for global optimization over continuous spaces. J. Glob. Optim. Int 11, 341–359 (1997)
4. Dorigo, M., Dicaro, G.: Ant colony optimization: a new meta-heuristic. IEEE Int. Conf. Evol. Comput. 2, 1470–1477 (1999)
5. Wenlog, F., Johnston, M., Zhang, M.: Soft edge maps from edge detectors evolved by genetic programming. In: CEC 99, International Conference on Evolutionary Computation. 2, pp. 1–8 (2012)
6. Storn, R.: Differential evolution design of an IIR-filter. In: IEEE International Conference on Evolutionary Computation. pp. 268–273 (1996)
7. Geem, Z.W., Kim, J.H., Loganathan, G.V.: Harmony search optimization, application to pipe network design. Int. J. Model. Simul. 22, 125–133 (2002)
8. Geem, Z.W., Tseng, C., Park, Y.: Harmony search for generalized orienteering problem: best touring in China., Springer Lect. Notes in Comput. Sci. 3412, 741–750 (2005)

9. Hitoshi, I.: Using genetic programming to predict financial data. In: CEC 99, International Conference on Evolutionary Computation. pp. 244–251 (1999)
10. Parpinelli, R.S., Lopes, H.S., Freitas, A.A.: Data mining with an ant colony optimization algorithm. Springer Lect. Notes in Comput. Sci. 6, 312–332 (2002)
11. Thangaraj, R., Pant, M., Abraham, A., Bouvry, P.: Particle swarm optimization: hybridization perspectives and experimental illustrations. Appl. Math. Comput. **217**, 5208–5226 (2011)
12. Mahdavi, M., Fedanghary, M., Damangir, E.: An improved harmony search algorithm for solving optimization problems. Appl. Math. Comput. 1567–1579 (2007)
13. Ghosh, A., Das, S., Chowdhury, A., Giri, R.: An improved differential evolution algorithm with fitness-based adaptation of the control parameters. Inf. Sci. 3749–3765 (2011)
14. Ali, M.M., Kaelo, P.: Particle swarm optimization: particle swarm optimization for global optimization. Appl. Math. Comput. **196**, 578–593 (2008)
15. Shi, X.H., Liang, Y.C., Wang, L.M.: An improved GA and novel PSO-GA-based hybrid algorithm. Inf. Process. Lett. **93**, 255–261 (2005)
16. Holden, N., Freitas, A.A.: A hybrid particle swarm/ant colony algorithm for the classification of hierarchical biological data. In: SIS 2005, International Symposium on Swarm Intelligence. pp. 100–107 (2005)
17. Esmin A.A.A., Torres, G.T., Alvarenga, G.B.: Hybrid evolutionary algorithm based on PSO and GA mutation. In: The proceedings of the sixth International Conference on Hybrid Intelligent Systems, pp. 57 (2006)
18. Li, H., Li, H.: A novel hybrid particle swarm optimization algorithm combined with harmony search for high dimensional optimization problems. In: The International Conference on Intelligent Pervasive Computing. pp. 94–97 (2007)
19. Ciornei, I., Kyriakides, E.: Hybrid ant colony-genetic algorithm (GAAPI) for global continuous optimization. IIEEE Trans. Syst., Man, Cybern. - Part B: Cybern. **42**, 234–245 (2012)
20. Ulker, E.D., Haydar, A.: Comparing the robustness of evolutionary algorithms on the basis of benchmark functions. Adv. Elect. Comp. Eng. **13**, 59–64 (2013)

# Author Index

© Springer International Publishing Switzerland 2015
S. Fidanova (ed.), *Recent Advances in Computational Optimization*,
Studies in Computational Intelligence 580, DOI 10.1007/978-3-319-12631-9

Printed by Printforce, the Netherlands